PESTICIDES
Problems, Improvements, Alternatives

PESTICIDES
Problems, Improvements, Alternatives

Edited by

Frank den Hond, Peter Groenewegen and
Nico M. van Straalen

Blackwell
Science

© 2003 Blackwell Science Ltd,
a Blackwell Publishing Company
Editorial Offices:
9600 Garsington Road, Oxford OX4 2DQ, UK
  Tel: +44 (0)1865 776868
Blackwell Science, Inc., 350 Main Street,
Malden, MA 02148-5018, USA
  Tel: +1 781 388 8250
Iowa State Press, a Blackwell Publishing
Company, 2121 State Avenue, Ames, Iowa
50014-8300, USA
  Tel: +1 515 292 0140
Blackwell Publishing Asia Pty Ltd,
550 Swanston Street, Carlton South,
Victoria 3053, Australia
  Tel: +61 (0)3 9347 0300
Blackwell Wissenschafts Verlag,
Kurfürstendamm 57, 10707 Berlin, Germany
  Tel: +49 (0)30 32 79 060

First published 2003 by Blackwell Science Ltd

Library of Congress
Cataloging-in-Publication Data
is available

ISBN 0-632-05659-2

A Catalogue record for this title is available
from the British Library

Set in 10/13pt Times
by Bookcraft Limited, Stroud, Gloucestershire
Printed and bound in Great Britain by
MPG Books, Bodmin, Cornwall

For further information on
Blackwell Science, visit our website:
www.blackwellpublishing.com

Contents

Contributors

David Barling is a senior lecturer in Food Policy at the Institute of Health Sciences at City University, London EC1 0HB. He has published on aspects of the governance of food and agriculture, notably the regulation of genetic modification throughout the food chain. He is member of the joint Sustain and the UK Food Group working group on agriculture and trade policy. (Sustain and the UK Food Group are two umbrella organisations representing over 100 food and agriculture and development NGOs.)

Susan Carr is a senior lecturer at the Open University in the UK, writing on systems management, farmland conservation management, environmental decision making and ethics. She has co-ordinated a series of research projects on the regulation of genetically modified crops in the European Union, with funding from the EU and from the UK's Economic and Social Research Council. She has been a member of a government Foresight working group on sustainability indicators for the food chain. She can be contacted at the Open University, Faculty of Technology, Milton Keynes MK7 6AA, UK.

Harrie A.J. Govers works at the University of Amsterdam, Department of Environmental and Toxicological Chemistry, Institute for Biodiversity and Ecosystem Dynamics, Nieuwe Achtergracht 166, 1018 WV Amsterdam, Netherlands. He has been a full-time professor of environmental chemistry since 1988. He has published widely on the physical, theoretical, analytical and biochemistry of organic micropollutants and their environmental fate. His special interest is the prediction of environmental chemistry properties required for fate models.

Nicolien M. van der Grijp is a researcher at the Institute for Environmental Studies. After having completed her studies in Dutch Law, she completed the Academic Professional Training in Environmental Management (UBM). Her current research work focuses on two distinct fields: the implementation of environmental policy of the European Union, and the mechanisms that influence environmental performance within national and international product chains, especially in the food sector. She can be contacted at Vrije Universiteit, Institute for Environmental Studies, De Boelelaan 1087, 1081 HV Amsterdam, Netherlands.

Peter Groenewegen is associate professor of organisation theory. He has done research in technology assessment in crop protection and materials technology. He has published on corporate environmental strategy and innovation networks. He can be contacted at Vrije Universiteit, Faculty of Social and Cultural Sciences, Department of Public Administration and Communication Science, De Boelelaan 1081c, 1081 HV Amsterdam, Netherlands.

Frank den Hond is assistant professor of competitive and corporate strategy in a recently established programme on Policy, Communication and Organisation. He has published on corporate environmental strategy and innovation in the automotive and agrochemical industries. He can be contacted at the Vrije Universiteit, Faculty of Social and Cultural Sciences, Department of Public Administration and Communication Science, De Boelelaan 1081c, 1081 HV Amsterdam, Netherlands.

Alan Irwin is professor of sociology in the Department of Human Sciences at Brunel University, West London. His books include *Sociology and the Environment* (Polity Press 2001) and *Citizen Science* (Routledge 1995). He was co-editor (with Brian Wynne) of *Misunderstanding Science?* (Cambridge University Press). Alan Irwin has research interests in environmental sociology, public understanding of science, and science and technology policy. He can be contacted at the Faculty of Arts and Social Sciences, Brunel University, Uxbridge, Middlesex UB8 3PH, UK.

Tanja de Koeijer is with the Nature Policy Assessment Office, PO Box 47, 6700 AA Wageningen, Netherlands. Her special interest concerns the economic aspects of wildlife conservation by farmers. She received her PhD in Agricultural and Environmental Sciences from production Wageningen University, Netherlands.

Martin J. Kropff graduated in 1976 at the State University of Utrecht and obtained his PhD in 1989 at Wageningen University. From 1984 onwards he was assistant professor, working on crop–weed interactions and modelling the effects of air pollutants on crops. During 1991–95 he was senior agroecologist at the International Rice Research Institute in The Philippines where he was active in the field of crop modelling and weed ecology. Since 1995 he has been full professor at the Wageningen University on crop ecology and weed science and since 1997 also director of the graduate school for production ecology and resource conservation. He is currently Director-General of the Division of Plant Sciences of Wageningen University and Research Centre, president of the European Weed Research Society and editor of *Weed Research* and *Agricultural Systems*. He has (co-)authored over 80 publications in international journals and over ten books. Martin Kropff can be contacted at Wageningen University, Department of Plant Sciences, Crop and Weed Ecology Group, Haarweg 333, 6709 RZ Wageningen, Netherlands.

Onno Kwast is at the University of Amsterdam, Department of Environmental and Toxicological Chemistry, Institute for Biodiversity and Ecosystem Dynamics, Nieuwe Achtergracht 166, 1018 WV Amsterdam, Netherlands. He participated as a master student of environmental chemistry.

Pim Leonards is at the Netherlands Institute for Fisheries Research (RIVO), PO Box 68, 1970 AB IJmuiden, Netherlands. He published his PhD thesis in the area of the analysis, food chain transfer and critical levels of PCBs in mustelids.

Kevin Parris has been working as an economist at the Environment Division of the OECD Agriculture Directorate, 2 rue André Pascal, 75016 Paris, France, since 1984, on policy analysis and developing agri-environmental indicators. He received his education at the University of Reading and obtained his PhD in agricultural economics. He worked previously at the Universities of Aberystwyth, Oxford and Reading; the UK Foreign Office; UNCTAD; FAO; and on secondment from OECD with the Australian Bureau of Agricultural and Resource Economics, ACT, Australia.

André van Roon is at the University of Amsterdam, Department of Environmental and Toxicological Chemistry, Institute for Biodiversity and Ecosystem Dynamics, Nieuwe Achtergracht 166, 1018 WV Amsterdam, Netherlands. His special interest concerns the risk evaluation and prediction of biodegradation rate constants of organic pesticides originating from the biogenesis by plants.

Henry Rothstein is at the London School of Economics, Centre for Analysis of Risk and Regulation, Houghton Street, London WC2A 2AE, UK. His main research interests are on risk regulation regimes and in particular the role of public opinion, the media, interest groups and regulatory professionals in shaping risk policy. Recent publications include *The Government of Risk* (Oxford University Press 2001) with Christopher Hood and Robert Baldwin, and articles in journals including *Public Law, Administration and Society*, and *Health, Risk and Society*.

Geert R. de Snoo is at Leiden University, Centre of Environmental Science, PO Box 9518, 2300 RA Leiden, Netherlands. He is associate professor and head of the section Ecosystems and Environmental Quality. He has published widely on aspects of integrating biodiversity (flora and fauna), environment (pesticides) and agriculture (costs and perceptions).

Nico M. van Straalen is professor of animal ecology at the Vrije Universiteit, Amsterdam, Faculty of Earth and Life Sciences, De Boelelaan 1085, 1081 HV Amsterdam, Netherlands, where he teaches Evolutionary Biology, Ecology and Ecotoxicology. He is interested in ecological risk assessment of soil pollution and in particular the mechanisms of genetic adaptation to heavy metals in springtails, a group of primitive insects. He supervises a research group on ecology of soil invertebrates, in which projects are conducted on soil function, nutrient cycling, effects of heavy metals and the relationship between biodiversity and community structure. He is also interested in studies of communication between science and policy, especially regarding risk assessment techniques using species sensitivity distributions.

Paul C. Struik (1954) graduated in 1978 and obtained his PhD in 1983, both at Wageningen University. From 1982 onwards he was associate professor of Wageningen University, working on physiology of tuberisation and tuber growth in potato. Since September 1986 he has been full professor in crop and grassland science. Currently he has a chair in crop physiology. He is secretary/treasurer of the European Association for Potato Research, co-ordinating editor of *Potato Research* and the *Netherlands Journal of Agricultural Science*, and editor of *Annals of Applied Biology*. Since 1982, he has (co-)authored over 120 publications in international journals and several books. Paul Struik can be contacted at Wageningen University, Department of Plant Sciences, Crop and Weed Ecology Group, Haarweg 333, 6709 RZ Wageningen, Netherlands.

Scott M. Swinton is associate professor in the Department of Agricultural Economics at Michigan State University, East Lansing, MI 48824–1039, USA. His research focuses on agricultural production economics and environmental management. He has published widely on the economics of technologies and policies related to pest management and precision agriculture.

Elizabeth M. Vogelezang-Stoute works as a researcher at the Centre for Environmental Law of the University of Amsterdam, Postbus 1030, 1000 BA Amsterdam, Netherlands. She specialises in pesticides law and is writing her dissertation, which is a comparative law study on the authorisation of pesticides. This study is a continuation of an earlier research project she did at the Centre for Environmental Law, together with Chemistry Sciences. She has published widely on European Community and national (Dutch) pesticides law and annotates case law on this subject.

Yukio Yokoi is deputy director in the Fruits and Flower Division of the Ministry of Agriculture, Forestry and Fisheries, 1–2–1, Kasumigaseki, Chiyoda-ku, Tokyo, Japan. He received his education at Cornell University (MPS in international agriculture and rural development) and Tokyo University (BS in agricultural chemistry). He worked previously in the OECD Agriculture Directorate and the Environment Agency of Japan.

Pim de Voogt is associate professor at the University of Amsterdam, Department of Environmental and Toxicological Chemistry, Institute for Biodiversity and Ecosystem Dynamics, Nieuwe Achtergracht 166, 1018 WV Amsterdam, Netherlands. His research interests are environmental chemistry of persistent organic compounds and their transformation products, structure–activity relationships and development of methods for their identification and quantification.

Ada Wossink is an associate professor in the department of Agricultural & Resource Economics at North Carolina State University, Department of Agricultural & Resource Economics, Raleigh, NC 27695–8109, USA. She received her PhD in Agricultural and Environmental Sciences from Wageningen Agricultural University in The Netherlands and is a Fulbright Alumnus. Her research interest is in the economics of production and environmental management in agriculture.

Acknowledgements

We wish to thank the Vrije Universiteit for having subsidised this project generously (USF grant no. 95/19, 'Milieu-effecten en maatschappelijke implicaties van "getransformeerde micro-verontreinigingen" '); NWO, the Netherlands Organisation for Scientific Research, for financial support; Désirée Hoonhout for her kind assistance in editing various chapters; the authors for having persevered with us in finalising this volume; and the reviewers to the various chapters, listed opposite, for their prompt and constructive reviews.

List of Reviewers

The editors and authors are very grateful for the kind review work and the constructive suggestions that were made by the reviewers listed below.

Gerald Assouline Université Pierre Mendès France de Grenoble 2, Grenoble, France

Henry Buller Cheltenham and Gloucester College, Cheltenham, UK

Susan Carr Open University, Milton Keynes, UK

Cees Gooijer Vrije Universiteit, Amsterdam, Netherlands

Freek de Meere Vrije Universiteit, Amsterdam, Netherlands

Willem Halffman Universiteit van Amsterdam, Netherlands

Vicky Hird Sustain, London, UK

Jan Linders RIVM, Bilthoven, Netherlands

Jamie Morrison Imperial College at Wye, Ashford, UK

Jesper Schou Danish National Environmental Research Institute, Roskilde, Denmark

Hanna Sevenster Universiteit of Amsterdam, Amsterdam, Netherlands

Geoff Squire Scottish Crop Research Institute, Invergowrie, Dundee, UK

Joyce Tait University of Edinburgh, Edinburgh, UK

Bill Vorley International Institute for Environment and Development, London, UK

Lori J. Wiles Colorado State University, USA

Ada Wossink North Carolina State University, Raleigh NC, USA

Preface

Since World War II, it has been impossible to imagine agriculture without the use of pesticides. It has served the economic interests of farmers and chemical industries alike, and facilitated the pursuit of rising levels of agricultural production at relatively low (private) cost. Generally, agricultural policies have been conducive to the use of pesticides, at least until recently.

However, after more than 50 years of pesticides use, the perceived blessings have faded, and for quite some observers, even turned into a curse. It is certain, though, that the ignorance and negligence that long have accompanied the development and use of pesticides have vanished. Change is in the air. But what direction should this change take?

This book seeks to answer this question. It provides us with a balanced collection of papers that addresses different angles, views and strategies. Due account is taken of institutional aspects, the dynamics of technological development and pressing policy issues, especially at international levels.

In other words, this book provides us with an intellectual framework with which to think about the problems, improvements and alternatives of pesticides use in agriculture. But the reader has to find his own way. Ultimately, it is a search for a sustainable agriculture that can feed a growing world population. And sustainability with respect to agriculture comprises such diverse objectives as health, food, habitats, biodiversity and landscapes, in addition to adequate income levels for farmers, not only in Western countries but especially so in the developing world.

Indeed, quite a task and a great challenge.

Harmen Verbruggen
Professor of International Environmental Economics
and Director of the Institute of Environmental Studies,
Vrije Universiteit, Amsterdam

Chapter 1
Questions Around the Persistence of the Pesticide Problem

Frank den Hond, Peter Groenewegen and Nico M. van Straalen

Introduction

Sustainable agriculture includes sustainable pest management. But how sustainable are current pest management practices that rely heavily on the use of pesticide products? This question has provoked considerable, often heated, debate. Positions in the debate have become entrenched, both on the side of those who believe that the (controlled) use of pesticide products can and does contribute to sustainable agriculture, and on the side of those who believe the opposite (Hamlin & Shepard 1993). The former point out that farmers who use pesticide products and other benefits of modern agricultural technology, such as fertilisers and seeds, are able to realise higher yields on their lands. Higher yields enable the feeding of a growing world population, and reserves land for uses such as recreation and nature conservation that otherwise would need to have been turned into productive agricultural land (Oerke *et al.* 1994). Their opponents argue that intensive agricultural use of land has led to serious environmental degradation while not providing a solution to hunger (e.g. Dingham 1993; Vorley & Keeney 1998). Moreover, they argue, by adopting intensive agricultural production systems, farmers lock themselves into a spiral of debts and further loans so as to invest in additional productivity growth. We do not wish to step into such a debate by choosing sides. It is our conviction that there is some truth on both sides. Much of the debate is unproductive black-and-white schematising. The greys have disappeared. Our ambition in this book is to bring the greys back into the debate. This volume aims to analyse the question of why the environmental impact of pesticide use continues to be of serious concern – the persistence of the pesticide problem – as well as to identify and assess critically various improvements and alternative strategies to circumvent the current problems of agricultural pesticide use.

This introductory chapter continues explaining the motives behind our questions and ambitions, and developing a conceptual framework of three interrelated spheres – agricultural production, socio-economic institutions, and agricultural innovation – that has guided our selection and direction of the contributions in this volume. The chapters are introduced in the final section of this chapter.

Are environmental problems of pesticides still an issue?

While it is obvious that there are problems related to the past and current use of pesticide products, our question is: what is the origin and structure of the problem? One might expect that over time individual and institutional learning in three areas should have resulted in significant improvements: (1) policies regulating the use of pesticide products, (2) the operational use of pesticide products, and (3) science-based innovation, including the development of new active substances, formulations, and application technologies, as well as alternatives to chemical crop protection. If, at least in the Western, industrialised world, pesticide products and their agricultural use have been regulated for at least half a century, if farmers have built up, individually and collectively, an accumulated experience with agricultural pesticide use over a period of anywhere between 50 and 150 years, depending on the geographical region, and, if for over a century, industry has been trying to profit by marketing more effective, cost-efficient and safer pesticide products, why then are Western, industrialised societies still facing problems related to pesticide use?

Our approach is limited to pesticide use in Western, industrialised countries, which is not to downplay the adverse impacts of pesticide use in developing countries. On the contrary, we would agree with many observers that the pesticide problem is far worse in those countries, for example, because of lower levels of education, illiteracy, lack of functioning regulatory systems, lack of money for protective measures, and the availability of pesticide products forbidden in Western countries. However, our analysis of the pesticide problem in Western, industrialised countries is directed to a different end. We want to analyse the manner in which the differences in context of various elements of the pesticide problem structure the societal and individual learning of actors. Such learning experiences and the solution to partial problems of pesticide use have structured different but intertwined development paths.

Pesticide policies have always had a strong focus on regulating negative externalities. Early policies aimed at ensuring the efficacy of pesticides in killing or deterring undesired organisms. Pesticide producers were only allowed to sell products with proven efficacy. Early regulation was thus directed at ensuring effective transactions and avoidance of non-effective products. Later, additional policy objectives were introduced, including the protection of occupational health and safety (protection of farmers and husbandry) and food residues (protection of consumers). Since the early 1960s, when the negative environmental impacts of pesticide use became a topic of societal debate (Carson 1962; Briejèr 1967), an increasingly refined and detailed regime of measures was implemented in order to reduce environmental impacts of pesticide use. In this context, the implementation of EU Directive 91/414 and its annexes is to be considered a landmark. Pesticide policies thus focus on regulatory design to curb negative effects, notably by implementing maximum acceptable concentrations (norms) and threshold concentrations for individual compounds in specific environments (food, water, soil and air) and on pesticide application routines and techniques. Several positive changes have been achieved, including bans on various pesticides; outstanding are bans on DDT and the so-called 'drins', and more recently on the 'dirty dozen' POPs, the persistent organic pollutants.

The rationale of regulating the negative externalities of pesticide use by authorising their use – or even the conditions of their use – is based on a strong belief in the possibility of prediction by various experimental methods and modelling the environmental fate of pesticide products. Today, for example, a pesticide company is permitted to market a pesticide product for a certain period of time, if it proves – with state-of-the-art science and technology – that the product is 'safe'. In this context 'safe' means that, under the assumption of proper use by the farmer, the product is not expected to result in unacceptable environmental degradation. To this end, a set of criteria based on compound characteristics has been legally established including criteria on degradation, mobility and toxicity. Such an authorisation system implies the extensive generation and modelling of relevant data from laboratory and field tests under controlled conditions prior to submission of the authorisation request to the relevant authorities.

Farmers, too, have developed experience with pesticide products. Most farmers know about the hazardous nature of pesticide products and their handling, in some cases caused by unfortunate accidents, more often because they have read label instructions or listened to the advice of extension services or agrochemical sales advisors. Undoubtedly, neighbouring farmers and colleagues in farmers' study groups will be very anxious to hear about experiences with newly introduced pesticide products, especially about the product's efficacy in pest control, its cost and ease of handling. Learning and education with regard to use has been based on the experience as well as the marketing arguments of the agrochemical industry. Negative consequences have also been encountered from the interaction between methods and results. Thus widespread occurrence of resistance to chemicals on the one hand has certainly played a role in the growth of doubts within the farming community and a turn towards alternative methods. On the other hand it is one of the reasons why many farming organisations argue for a variety of chemicals and are disappointed at the slow introduction of new compounds.

Many new products have been introduced since the Bordeaux mixture was first applied about 150 years ago. Several substituted for older, less effective or more risky products. Modern pesticides show little resemblance to the by-products of the early chemical industry – pharmaceutical products, synthetic dyes, fibres and plastics – that were used as pesticide products. Today's pesticide products are high-tech products: consider the systemic character of many fungicides, the low dose rates of several herbicides and the specificity of insecticides. They exhibit higher selectivity and reduced persistence by the introduction of new active ingredients and formulations. The agrochemical industry spends huge sums, not so much on the identification of new active ingredients as on the testing for environmental, human and eco-toxicological characteristics of those compounds and on the quest for the most effective formulation. Additionally, many developments have taken place in application technology. Early application technologies were often based on the spraying of pesticide products from back-packed containers under air pressure. Mechanisation of spraying (larger containers, tractor- or plane-based), evolving spraying techniques (finer droplets, nozzle orientation) and improvements of formulation technology (seed coatings, water soluble granules, adhesives, flow compounds) are examples of further developments that have taken place in the introduction of more effective application technologies. Containers have improved to make handling safer. In several countries, including The Netherlands, plane sprayings are restricted, and spray machinery has to be controlled annually in order to get the pesticide product where it should be, that is, on the crop in the field, rather than drifting away into surrounding waterways, other fields, or non-agricultural areas.

Despite the evolution of the regulatory framework, the development of new agricultural production techniques and the refinement of the technical arsenal, in many industrialised countries there continues to be serious concern about the adverse impacts of agricultural pesticide use. Various problems related to pesticide use appear to be rather resistant. Fine-tuning of the system has resolved some problems, but has also resulted in the emergence of new problems that may require a more fundamental reshaping of the system that has been built around the agricultural use of pesticide products. Simultaneously, incomplete knowledge of existing and potential effects may influence our perception of the dangers and benefits of pesticide use. Many examples to support our claims could be given; much has been published about the impacts, risks and dangers of pesticide use (e.g. Vorley & Keeney 1998; Dingham 1993; de Snoo & de Jong 1999; Hough 1998). Here is a selection of current issues.

(1)   Pesticides and their metabolites can be found everywhere: in fresh water, groundwater, soil, food and even in faraway oceans. Some of the pesticides found have long been banned. In other instances, the residues found on food were of a pesticide not allowed for that particular application. Moreover, in many cases the concentration in which pesticides are found exceeds the established norms and threshold values, at times by factors of ten to a hundred (de Snoo & de Jong 1999). Some pesticides were found to have effects on the endocrine system and others proved to be carcinogenic. Several pesticide residues were found to accumulate in human and biological food chains. Although the days are past when birds fell dead from the sky, predator populations are still threatened by bioaccumulation of pesticide residues through food chains. Although in most OECD (Organization for Economic Cooperation and Development) countries there are effective food safety policies in place, this is generally not so in many developing countries.

(2)   The detection of persistent organochlorines in ecosystems far away from any industrial or agricultural source has raised concern about the possible ecological effects of these compounds in ecosystems of a pristine nature. Pesticides that are banned from most industrialised countries are still measured in remote regions. The question is whether present registration procedures are sufficiently conservative to avoid possible ecological effects of residues transported to remote areas. The reasons to be especially concerned about effects following long-range transport are: (a) once a pesticide becomes airborne, it is out of control, (b) pesticide residues are transported over large areas, (c) environmental conditions at remote ecosystems may promote long residence times and (d) organisms in remote ecosystems may be more vulnerable than agro-ecosystems (van Dijk *et al.* 1999).

(3) Pesticides are commonly evaluated using a limited number of simple ecotoxicity tests as indicators for ecological effects. There is serious doubt whether such tests (usually short-term exposure to relatively high dose rates of a single species under artificial conditions) can be indicative of effects in natural systems in which trophic interactions, indirect effects and secondary poisoning may operate. Relatively little is known, too, about how ecosystems may recover from pesticide impacts and how the rate of recovery can be included in the risk assessment. Mesocosm studies show that the primary effects of pesticides can be well predicted from the laboratory experiments, provided that the exposure conditions in the mesocosm are measured. Secondary effects, however, are very difficult to predict, since these depend on the physical and ecological structure of the system, and many environmental parameters (Hill *et al.* 1994; van Straalen & van Rijn 1998).

(4) A rapidly increasing number of species have developed resistance to the action of individual as well as groups of chemically related active ingredients (Weber 1994). Due to the dynamic responses of local ecosystems to pesticide use (and other agricultural measures), such as the build-up of resistance and the occurrence of secondary plagues, attempts to control the side effects of pesticide use look like hitting a moving target. In this respect, Zeneca scientists Clough and Godfrey observe that: 'As methods of crop production have become more intensive, the incidence and severity of fungal diseases have increased' (Clough & Godfrey 1995). The repeated use of the same pesticide over time leads to the evolutionary selection of those pest organisms that have developed resistance to a particular pesticide or family of pesticides. Accordingly, an increasing number of fungi, weeds and insects have become resistant to pesticide sprays. Indeed, some observers have noted a perverse effect of the general use of pesticides, namely that crop losses due to insect invasion have actually increased with increasing pesticide use (Dalzell 1994).

One cannot but conclude that, if sustainable agricultural practice is considered, the objective, current agricultural pesticide use and related policies fail in many respects. The question next to be addressed is: why is this the case?

A framework for analysis: production, innovation, institutions

To understand why pesticide problems are so persistent, what the reasons are for 'regulatory failure', and how these problems might be solved or circumvented, is a complex and challenging endeavour. It is further complicated by the facts that the pesticide debate has become highly politicised and that the positions of several participants in the debate have become rather entrenched. It is simplistic to one-sidedly blame either the agrochemical industry or farmers for all the problems and uncertainties. Indeed, a large number of often interrelated factors have forced farmers increasingly to rely on pesticide products (Pretty *et al.* 1998). In our view, a workable starting point for this task is to take a broader view. In order to understand pesticide use and the difficulty of controlling and regulating external effects of their use, one should consider a complex, dynamic and interactive system of agricultural production, agricultural innovation, and socio-economic institutions (Fig. 1.1). Broad lines of analysis should focus on these three subsystems or 'spheres', as well as their interaction. In the following we discuss the separate spheres as well as their interactions.

In the first sphere, agricultural production takes place – crop rotations are decided upon, investments made and seed sown, in the prospect of making a profit. Under conditions of uncertainty, farmers make *ex ante* decisions that impact on the level of income they might gain and the level of risk they wish to run. Pesticide products have become an important variable in such decision making, and hence in agricultural production. To some extent, farmers' decisions on pesticide use are guided by the products and services of pesticide producers and extension services, as well as structural characteristics of their business environment: regulations, prices and subsidies.

However, this sphere is not static; it has evolved over time. New technologies have become available for pest control. One particularly important setting is the chemical industry. In the public research laboratories, also, agronomic knowledge has provided new insights, partly in response to practical problems in agricultural production. Each of these innovative processes is denoted with the second sphere of agricultural innovation.

In order to complete this picture we argue that many of the processes in agricultural production and innovation have a direct relation to political and economic processes that take place in a wider context of socio-economic institutions: our third sphere. Production and innovation are confined and restricted by socio-economic institutions, but simultaneously, the latter develop in response to internal and external problems of agricultural production and innovation. The eventual success of suggestions for improvements upon, and alternatives to, current pesticide use depends on how well they are received within the context of current socio-economic institutions, or to what extent the latter may be changed to increase the viability of promising improvements and alternatives.

In order to set the scene for the type of interactions between the three spheres we will discuss briefly the interaction between agricultural production and the socio-economic sphere. Over the past century, agricultural production has changed dramatically (den Hond *et al.* 1999). Before World War II agricultural production was

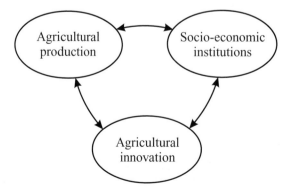

Fig. 1.1   A conceptual framework of three 'spheres'.

characterised by a relatively low input of capital, by high labour input, mixed farming and relatively low yields. For most crops, yields were restricted by local climatic, geological (soil) and ecological conditions, although mineral fertilisers were known and applied already in the second half of the nineteenth century. In the period of reconstruction just after World War II, political and socio-economic pressure changed the nature of agricultural production. Food security at low prices for a growing population, self-sufficiency in agricultural production and securing farmer income at acceptable levels became political priorities. These policy objectives translated into the maximisation of yields through the specialisation and rationalisation of agricultural production, by means of increasing scale, minimising labour input (for example, through increased mechanisation) and maximising the efficiency of various inputs, such as improved cultivars, better timing of farming operations, and management of soil moisture. The large-scale and low-cost availability of inputs such as pesticides was instrumental in minimising labour input. In this way, the traditional limits on agriculture were eliminated. In Western Europe, EEC and EU agricultural policies have significantly enhanced this development, notably by ensuring minimum prices and subsidising exports to world markets.

In certain respects, this policy was highly successful. The shift to intensive farming systems in the aftermath of World War II has allowed a break from dependence on crop rotations and livestock. For instance, world-wide cereals production has increased more than 2.5-fold, a growth rate higher than that of population growth (Brown 1994). Similarly, Vos (1992) reports a more than doubling of tuber yields in potato production in The Netherlands since 1950. Countries that adopted agricultural policies to stimulate this shift obviously enhanced the competitiveness of their agricultural exports.

But undesired consequences must be pointed out as well, which relate to (1) pressure on farm profitability and (2) increased vulnerability of the farming system in terms of sustainability. The processes of specialisation, rationalisation and technological innovation have advanced to the extent that farm profitability has been adversely affected. By the late 1980s, political concern became widespread that self-sufficiency had turned into overproduction, incurring high social cost. Farmers' incomes were under pressure from low prices for most agricultural crops. According to many observers, because of increased specialisation the agricultural production system had become vulnerable to calamities, such as unexpected outbreaks of pests and diseases, extreme weather events, and the development of pest resistance. Pest resistance causes reduced marginal returns to pesticide and fertiliser input, increased costs to the farmer, and increased public concern about the preservation of landscape and the pollution of natural habitats. It is understood that such effects are a consequence of the intensive cultivation of a small variety of crops in large areas of monoculture.

Current agricultural policy – for example the new European common agricultural policy – is aimed at the reduction of overproduction, liberalisation of world trade, and the protection of landscape and natural habitats. Farmers have to increase the quality of their output in terms of product appearance, delivery and nutritional value, while reducing the environmental risk of the decisions they take regarding crop management. Simultaneously, they have to control cost because of increased competition from the world market. This should be accomplished by, among other measures, a reduction of fertiliser and pesticide inputs and by the stimulation of more sustainable forms of agricultural production. Pesticides should be more effective at lower application doses, less toxic to non-target organisms, non-persistent and not pose a threat to groundwater quality. Such demands pose a serious innovation challenge to the agrochemical industry, traditionally a major supplier and innovator of pest control technology. One could speculate about the industry's willingness and capability in redressing its innovative activity to address such changing market and social demands, as yet another causal factor for the persistence of the pesticides problem.

Countries which are now trying to restructure their agriculture along ecological lines, with incentives for reduced dependence on chemical inputs, are facing the prospect of World Trade Organisation (WTO) action against trade barriers. For example, European Commission rulings against national bans on the imports and sales of Bt corn in Austria, Italy and Luxembourg may be partially understood from the fear of violating WTO/GATT free trade rules. And even today the shift to intensification continues as horticultural production is moving to where labour costs and climate allow lowest cost production and year-round availability on supermarket shelves. This has consequences for pesticide use, worker protection, externalities of long-distance transport, and suppression of local (seasonal) production. Thus, one may argue, the scope for pro-active pesticide policies is limited. Authorities are trapped in a set of administrative procedures which were designed in an earlier period, and which are limited in their concerns as exemplified by current registration procedures.

In our discussion, 'socio-economic institutions' thus comprise, but are not restricted to, the regulatory system around the introduction and application of pesticide products. The regulatory subsystem has also evolved enormously over time. On the one hand, evolution in the regulatory system can be seen as a reflection of societal concern over external effects of pesticide use (Howard 1940; Carson 1962), which depend on the products themselves and how they are being used in practice. Apparently, acceptance of external effects has steadily decreased over time, forcing the regulatory system to lower risk levels. On the other hand, science, notably analytical chemistry and systems ecology, has achieved considerable progress, allowing for earlier detection of increasingly more subtle external effects. Relative inertia in the agricultural production system may be caused by its slow response to regulation, or by the relative dynamics of the regulatory system imposing shifting objectives onto the agricultural system.

Further limitations to regulatory control of the environmental impacts and other externalities of agricultural pesticide use are likely to include their incompleteness and the constant need of renewal and revision, if only because potential externalities of new active ingredients and new formulations cannot be fully known beforehand. Additionally, pesticide authorisation and regulatory control is likely to be restricted as far as it has traditionally considered farmer behaviour as a constant factor, rather than as a variable; there are indications that farmers may apply pesticides in ways that differ significantly from label instructions, not to mention illegal use. Moreover, it is likely that the societal definition of what are undesirable externalities will continue to evolve, partly in relation to the further development of ever-refined analytical techniques to detect residues.

This conceptual framework of three spheres allowed us to raise broad questions about potential causes of the persistence of the pesticide problem. Why are pesticide products applied in agricultural production in the first place, and how serious is the pesticide problem? How responsive is agricultural production to regulations attempting to reduce external effects? To what extent are socio-economic institutions, among which the pesticide authorisation system is eminent, setting the proper conditions for desired farmer behaviour and more sustainable agricultural production? How effectively do innovations develop and diffuse into actual agricultural production activities? How tailored is agricultural innovation to solving internal and external problems in agricultural production, such as external effects of pesticide use? What are the more likely directions of innovation under the conditions set by prevailing socio-economic institutions?

The brief sketch provided here has been used as a guiding tool for the invited contributions in this book. We have asked the authors to submit a first paper that was discussed in a workshop held in June 2000 in Egmond aan Zee, The Netherlands. We used the workshop to strengthen the contributions, establish links between chapters and identify what approaches we needed to establish a more complete discussion. Subsequent rewriting and a round of comments by reviewers have resulted in the present volume. Not all of these questions will be covered by the chapters in this book, and we will return to the insights that can be generated by combining different expert views at the end of this volume in the concluding chapter.

Summary of chapter coverage

This volume is an attempt to bring together, confront and eventually integrate the insights of experts from various backgrounds including law, agronomics, agricultural economics, environmental sciences, ecological sciences, chemistry and social sciences; many of the chapters themselves have already an interdisciplinary character.

Struik & Kropff (Chapter 2) present their broad vision regarding the agricultural use of pesticides. Over the past century, an encompassing technology package has developed, aimed at the modernisation and rationalisation of agricultural production, which includes chemical fertilisers, hybrid seeds, sophisticated knowledge, and the replacement of labour with machines. During this time, the use of pesticides in agricultural production has, on the one hand, significantly contributed to increases in productivity by suppressing predator species that feed on crops (parasites, diseases, herbivores) and species that compete with crops for light, nutrients and water (weeds). On the other hand, external effects (environmental pollution, impacts on non-target organisms) and second-order effects (resistance) of pesticide use have developed to such an extent that the alleged agro-economic benefits can be seriously questioned and environmental and social sustainability is threatened. What is needed, according to Struik & Kropff, is a drastic reduction of the dependence on chemical inputs, and the amount of chemical inputs used in modern agriculture. This translates into a systems approach to agricultural production, in which pests, weeds, diseases and so on are controlled, suppressed or prevented by making use of ecological principles. Pest control decision making should consider strategic aspects, such as implications for the entire cropping system. The authors are optimistic about the steps already taken in improving relevant ecological and agronomic knowledge and the promises of further developments in this respect.

The Struik & Kropff chapter is well placed at the beginning of this volume, since other chapters specify and refine their analysis and make suggestions for the opportunities and conditions under which the identified improvements and alternatives might be realised. Chapters 3 to 7 primarily focus on the persistence of the pesticide problem; Chapters 8 to 13 primarily focus on improvements and alternatives.

Vogelezang-Stoute (Chapter 3) discusses the working of the admission and regulation system around pesticides from the perspective of juridical practices and procedures. Pesticides have to be authorised or registered before they can be sold on the market. Originally, authorisation was granted if the product was proven effective. Later on, additional criteria were added to the authorisation process, such as matters concerning occupational health, protection of consumers, protection of husbandry, protection of non-target organisms, and protection of groundwater reserves and the environment at large. This development can be seen as the administration's responses to increasing public and scientific concerns over a widening range of topics related to the external effects of continuing agricultural pesticide use. Vogelezang-Stoute considers the authorisation process from the point of view that current procedures may not be able to prevent the occurrence of adverse effects of pesticides on the environment. The harmonisation of pesticide regulations under the European Commission's Directive 91/414 is advancing improvements in national regulation and implementation. However, there remain bottlenecks, which relate to the need for updating earlier decisions on the authorisation of pesticides under consideration of more recent standards; the speed of development of new scientific insights, and their subsequent inclusion in authorisation standards, on the cumulative and aggregate exposure to pesticides; and on the lack of an operational elaboration of the so-called 'substitution principle', according to which the necessity of pesticide authorisation is considered against the availability of less risky chemical or even non-chemical alternatives. While her analysis is based on Dutch legal practice and its adaptation to Directive 91/414, the effects of harmonisation and the introduction of non-health considerations have a wider importance and clearly show the important role of design of regulation.

Den Hond (Chapter 4) addresses the role of innovation in the agrochemical industry. The agrochemical industry has been hampered by increasingly high costs of research and development (R&D) efforts regarding the introduction of new chemicals. As R&D costs are such a dominant factor, competition appears to be more focused on the introduction of new agrochemicals in the light of strategic positioning *vis-à-vis* competitors, rather than on product development in relation to societal needs. This argument is developed starting from a detailed analysis of the development of strobilurin fungicides, the elaboration of what appears to be the industry's technological paradigm, and a discussion of the drivers and radicality of agrochemical innovation. The relevance of the latest is not so much 'radicality' *per se*, but in what respect the innovations are radical. If, as Struik & Kropff suggest in their chapter, a radically new approach to crop protection is to be developed, one that is based on a systems approach in which the use of pesticides is one of a considerably expanded set of means for crop protection, to what extent is the agrochemical industry responsive to such demands? It would appear, then, that the industry has serious difficulty in moving in this direction.

Irwin & Rothstein (Chapter 5) provide an 'institutional' analysis on the role of science in establishing standards for environmental and health protection, e.g. at what concentration level a certain compound in some environmental compartment should be considered as undesired pollution, and in the compliance testing of new active ingredients and pesticide products, e.g. how to prove that under specified use conditions a pesticide product will not build up to such concentrations in some environmental compartment that standards for environmental protection are exceeded. Such regulatory systems are dependent on the input of science and scientists, but the functioning of such systems does not and cannot be solely 'scientific', given the important commercial consequences of the outcomes, prevailing scientific uncertainties regarding 'accuracy' and 'pedigree' of the outcomes (Funtowicz & Ravetz 1990), externally set deadlines for decision making, highly politicised decision environments and other non-scientific factors. To suggest that scientists are captured in such systems, to the extent that their results cannot be trusted in other ways than as the products of loyal servants to their paymasters, is as misplaced as to consider standard setting and compliance testing as mere technical issues that do not deserve further interest. While agrochemical companies increasingly have integrated anticipatory tests and analyses within their R&D activities, it is argued that active involvement of these companies in the EU regulatory system is also based on their ability to provide essential information to regulators. In such a situation, it is hardly surprising that Irwin & Rothstein observe a striking consensus among the parties involved that problems in the regulatory system do not lie in the safety reviews, but in processes of implementation and enforcement: e.g. in relation to illegal use and misuse of (banned) pesticide products. Another consequence of this close-knit community of experts is that it is hard for external parties to obtain access, which might be considered an impediment to the development of more broadly shared and transparent risk management strategies.

Both Wossink & de Koeijer (Chapter 6) and de Snoo (Chapter 7) discuss inefficiencies in agricultural production at the level of the individual farmer and the difficulties that exist in overcoming them. They argue that farmers must make *ex ante* decisions regarding pest control in conditions of uncertainty, e.g. regarding the weather and other factors that influence pest, disease and weed incidence and gravity of infestation. Because of greater pest mobility, risk averse behaviour, such as preventive spraying, is more likely to occur against insects and diseases than against weeds. Since, moreover, overuse of herbicides is more likely to affect the crop yield and quality than overuse of fungicides and insecticides, the stage is set for farmers to make non-optimal pest control decisions.

Wossink & de Koeijer argue theoretically, and show empirically for the case of nitrogen and herbicide input in a sample of sugarbeet producing farmers in

Flevoland, The Netherlands, that these farmers are more likely to be economically efficient than environmentally efficient, that improving technical efficiency is key in improving economic and environmental efficiency, and, moreover, that over the years inefficiencies tend to be located with the same subset of farmers. However, telling these farmers how inefficient they are and what they could do about it is not necessarily going to improve this situation, because the rational response of a farmer to increased insight into best practices – the production possibility frontier (PPF) – depends on various factors, including in what respect the farmer is inefficient, the form of the PPF, and the desired level of environmental quality. Their result poses a considerable challenge to designing policies that aim at increasing or at least maintaining farm profits and increasing environmental quality.

De Snoo specifies the significance of variations from average and maximum recommended dose rates in a number of agricultural and horticultural crops; the per hectare cost of pest control might be a good predictor of pesticide overuse. Similarly, there is significant variation in the environmental impact of pesticide use among farmers who grow the same crop in the same area. However, since there is no direct relationship between a pesticide's potency (as indicated by the recommended dose rate) and the environmental impact of its use (depending amongst others on compound characteristics but also on factors such as mode and condition of application), policy instruments such as levies and taxes on pesticide products should be designed skilfully in order to reduce environmental impact of pesticide use. Such considerations provide convincing arguments for de Snoo to advocate pesticide reduction policies at the level of individual farms that impact upon operational decision making around crop protection.

Parris & Yokoi (Chapter 8) review three main types of agri-environmental indicators that are being developed in OECD countries in the context of pesticide policies: pesticide use, risks of pesticide use and pest management. While there is a clear need for such indicators from a policy perspective, it is less clear what they should look like in order to be effective. Various conflicting demands have to be brought together, such as scientific accuracy *versus* simplicity in order to enhance comprehensibility to policymakers and other stakeholders, and relevance *versus* data availability. The latter, data availability, is a major problem for indicator development since in several OECD countries the only data available – if at all – are data of pesticide sales rather than use. The implication of this disenchanting argument is that in such countries one can only speculate about the impact of pesticide policies.

Govers and his co-authors (Chapter 9) start by arguing that current registration policies are flawed in several important respects, e.g. that the required tests focus on active ingredients without consideration of potential impurities or metabolites. Moreover, the monitoring of active ingredients and their metabolites in biotic and abiotic systems is not organised in such a way as to enable feedback into re-authorisation decision making. The integration of tests and testing information along such lines would increase the chance of proper safety testing and procedures. The uncertainty surrounding the complete picture of effects of pesticides requires a more flexible approach to the application of scientific insights. To this end, a 'safety net' approach is sketched that covers all life cycle stage of pesticides. The safety net is built around chemical reactivity, because this is a key property with respect to the prediction and detection of potential impacts of new active ingredients. The authors review the availability of methods enabling the prediction and detection of environmental occurrence and effects of pesticides, their impurities and metabolites. Two examples are given and analysed. The case of MCPA emphasises the importance of early prediction of the formation of unwanted side products (impurities) during production. The second case of monoterpenoids–potential substitutes for current pesticides–treats the prediction of biodegradation.

Swinton (Chapter 10) discusses the opportunity that exists to reduce the environmental impacts of agricultural pesticide use through the emerging agricultural practice of site-specific pest management (SSPM). Within fields, pest incidence may vary considerably, implying that not everywhere in the field do equal doses of pesticides need to be sprayed. SSPM aims at accomplishing precisely this task: spraying only where spraying is needed. The earlier mechanisation of agricultural production resulted in uniform treatment of entire fields, because it would be too costly, or very impractical indeed, to use the machinery (planes) on, for example, only a section of the field. The combination of recent developments in a number of technologies, including those of geographic information and positioning, as well as those of pesticide application and automated sensing, promises cost reductions and environmental benefits in applying pesticide products only where needed. However, SSPM technologies do not provide equal opportunities for each type of pest to be managed. Important gains can be made in the management of weeds and perhaps nematodes, but due to their mobility, insect pests and diseases are more difficult to manage in this manner. Another potential limitation in the deployment of such technologies may be the amount of initial investment which might be difficult for individual farmers to bear.

Carr (Chapter 11) analyses the claims and counter claims regarding the application of biotechnology in pesticide reduction strategies. Arguments in the discussion include economic benefits in terms of increased yields, reduced pesticide applications and increased farmer income *versus* assessment of environmental and health risks in terms of resistance development, transfer of genes to other, related species and exposure to unexpected toxins and allergens. Today, practical experience with genetically modified (GM) crops has been limited to a few crops (mainly herbicide-tolerant corn, soybean and cotton, and Bt corn and cotton) and a handful of cropping seasons, in restricted geographical areas in the United States. This limited availability of practical experience is a serious obstacle to a rigorous assessment of the robustness of the various claims on benefits and risks. The evidence so far would suggest that GM crops might be a tool in what Carr calls the 'environmental management' form of sustainable agriculture, that is, a form of agriculture as an integrated system in which management skills are used to minimise the impact of agricultural production on the environment. However, this requires many institutional provisos relating to the control of this technology. How these are to be fulfilled remains a topic for further debate.

Carr's approach to the biotechnology debate, in particular the potential risks and benefits, is very productive because of the scrutiny by which she analyses claims

and counter claims. As van Dommelen (1999) has suggested, this allows for un-
founded claims to be refuted, or refined in such a way that proponents and adversaries
may develop a common understanding of which issues and questions are crucial for
decision making about biotechnology.

Van der Grijp (Chapter 12) signals that currently a number of constituencies,
outside of the traditional range of government authorities concerned with agricultural
production, food safety and environmental quality, have organised themselves so as
to influence pesticide use in agricultural production. Notably supermarkets and food
processing companies feel a strategic need better to control the conditions of agricultural production. Consumers have become more critical of food safety issues,
particularly in the aftermath of the European BSE scare and other outbreaks of
diseases in livestock, but the issues also include concerns over consumer health
and safety of GM food and pesticide residues. Supermarkets and food processors
have responded to these issues by implementing policies that are aimed at invoking
consumer trust in their brands, rather than in the food production system at large, as
has been the objective of food authorities in various countries. The policies imply
increased control over the precise conditions of, and operational decision making in,
agricultural production. They have done so successfully, it would appear, since the
spectacular growth of sales of organic products started when supermarkets created
space on their shelves for such products.

Barling (Chapter 13) addresses the question of how international agricultural policies, such as the Common Agricultural Policy (CAP) and its recent reforms in the
European Union, and international trade agreements, such as developed within the
context of the WTO, may impact on the development of and prospects for organic
farming as one example of an agricultural production system that provides a clear
alternative to the heavy use of pesticide products in conventional agriculture. Organic
farming, but also other labels used for alternatives in agricultural practice, need to be
assessed as a reaction to various aspects of current farming that forced farmers to
change both their methods and scale of operation. Organic farmers have benefited
from state support, but the consequential growth of their movement has had the effect
of institutionalising organic production practices. In this respect, has organic
farming, as a producer of social goods – such as biodiversity, the sustaining of natural
habitats, landscapes, and local communities – lost out to organic farming as a
productionist, yet extensified (because pesticide-free) alternative to conventional agricultural production systems? Paradoxically, the future for organic farming depends
not only on the movement's own appraisal of this shift in orientation, but also on the
question of whether necessary state support for organic farming is considered as
contributing to the production of social goods (which is expected to be allowed) or to
direct or indirect farmer income or production subsidies in the productionist mode,
which is expected to be increasingly difficult under international regulations.

References

Briejèr, C.J. (1967) *Zilveren sluiers en verborgen gevaren: Chemische preparaten die het leven bedreigen.* Sijthoff, Leiden.

Brown, L.R. (1994) Facing food insecurity. In: *State of the World 1994* (ed. L.R. Brown), pp. 177–97. Earthscan, London.

Carson, R. (1962) *Silent Spring.* Houghton Mifflin, Boston.

Clough, J.M. & Godfrey, C.R.A. (1995) Growing hopes. *Chemistry in Britain* (June), 466–69.

Dalzell, J.M. (1994) *Food Industry and the Environment: Practical Issues and Cost Implications.* Blackie, London.

van Dijk, H.F.G., van Pul, W.A.J., & de Voogt, P. (1999) *Fate of Pesticides in the Atmosphere: Implications for Risk Assessment.* Kluwer Academic Publishers, Dordrecht.

van Dommelen, A. (1999) *Hazard Identification of Agricultural Biotechnology: Finding Relevant Questions.* International Books, Utrecht.

Dingham, B. (1993) *The Pesticide Hazard.* Zed Books, London.

Funtowicz, S. & Ravetz, J.(1990) *Uncertainty and Quality in Science for Policy.* Kluwer Academic Publishers, Dordrecht.

Hamlin, C. & Shepard, P.T. (1993) *Deep Disagreement in U.S. Agriculture.* Westview Press, Boulder.

Hill, I.R., Heimbach, F., Leeuwangh, P. & Matthiesen, P. (1994) *Freshwater Field Tests for Hazard Assessment of Chemicals.* Lewis Publishers, Boca Raton.

den Hond, F., Groenewegen, P. & Vorley, W.T. (1999) Globalization of pesticide innovation and the locality of sustainable agriculture. *American Journal of Alternative Agriculture*, 14 (2), 50–58.

Hough, P. (1998) *The Global Politics of Pesticides.* Earthscan, London.

Howard, A. (1940) *An Agricultural Testament.* OUP, London.

Oerke, E-C., Dehne, H-W., Schönbeck, F. & Weber, A. (1994) *Crop Production and Crop Protection: Estimated Losses in Major Food and Cash Crops.* Elsevier, Amsterdam.

Pretty, J., Vorley, W.T. & Keeney, D. (1998) Pesticides in world agriculture: Causes, consequences and alternative courses. In: *Bugs in the system* (eds W.T. Vorley & D. Keeney), pp. 17–49. Earthscan, London.

de Snoo, G. & de Jong, F. (1999) *Bestrijdingsmiddelen en milieu.* Jan van Arkel, Utrecht.

van Straalen, N.M. & van Rijn, J.P. (1998) Ecotoxicological risk assessment of soil fauna recovery from pesticide application. *Reviews of Environmental Contamination and Toxicology*, 154, 83–141.

Vorley, W.T. & Keeney, D. (1998) *Bugs in the System.* Earthscan, London.

Vos, J. (1992) A case history: Hundred years of potato production in Europe with special reference to The Netherlands. *American Potato Journal*, 69, 731–51.

Weber, P. (1994) Resistance to pesticides growing. In: *Vital Signs 1994* (eds L.R. Brown, H. Kane & D.M. Roodman). W.W. Norton, New York.

Chapter 2
An Agricultural Vision

Paul C. Struik and Martin J. Kropff

Introduction

Agriculture will always have to cope with crop losses caused by biotic stresses. During the last fifty years these stresses were controlled by pesticides, a strategy that has proved unsustainable in the long run. New, ecological approaches will be system-oriented and supported by a thorough insight in the short-term, mid-term and long-term effects of control strategies on the economic sustainability of intensive farming, the effect on food security and the environmental impact. Such approaches should be based on prevention, optimal decision making and effective control strategies. For bacterial diseases prevention is possible, but control is still difficult to achieve with any method. For airborne fungal diseases, prevention (except through durable resistance) is virtually impossible and non-chemical alternatives for control are still not effective enough. For other pests, pathogens and weeds, non-chemical control strategies are gaining ground. The chemical era of crop protection may come to an end if a tremendous increase is realised in ecological and agronomic knowledge on how to prevent and control pests, diseases and weeds, based on insight into the dynamics of their occurrence, the factors controlling their behaviour and the risks of economic damage. Novel techniques are under development that will improve a crop's resistance and tolerance, will increase efficacy of control and will help to predict the chances of yield loss and assist in decision making on control measures. Nevertheless, more can be done to upgrade our ecological understanding at higher levels of aggregation and on transferring this knowledge to the individual grower.

The agronomic setting

During the last century a greater part of the increase in productivity per unit area and unit of labour has been achieved by the successful introduction of the use of chemical ingredients (either chemical fertilisers, chemical crop protectants or growth regulators), and the adaptation of crops to the use of these chemicals (e.g. Porceddu & Rabbinge 1997; Neumann 1997). In particular, the coherent use of a technology package containing a balanced input of both chemical fertilisers and chemical crop protectants has been successful (de Wit 1992). This success can be illustrated by the following examples. Around 1890, it took in The Netherlands 300 man-hours of labour to produce 1.5 tons of wheat per hectare; about a century later, this labour input was reduced to 15 man-hours per hectare, with an average yield of 8.5 t/ha (de Wit *et al.* 1987). In Italy, wheat yields increased from about one ton per hectare at the beginning of the twentieth century, absorbing about 70 man-days. At the end of the twentieth century wheat yields were four times as great and one hectare of wheat cultivation only absorbed about four man-days (30 man-hours) of labour (Porceddu & Rabbinge 1997). In the industrialised world such a development has taken place with all major crops. Once agriculture starts to profit from the rapid development of the life sciences, this process may even be accelerated (Sonnewald & Herbers 2001).

Crop production is unfeasible without some form of crop protection. Crop protection relies on the manipulation of the pest or disease (either chemically or non-chemically), the crop (such as through breeding or agronomic measures) or the environment (e.g. through intermediate host elimination or introduction of antagonists). Crop protection starts with protecting the propagules from which the crop is grown, during their production (e.g. by haulm killing), during storage, during preparation for use and immediately after sowing or planting (e.g. by seed coating). Crop protection is intensified when the crop is in the field and exposed to a diverse complex of biotic stresses, and continues throughout the growing season and even beyond that (with protection of produce stored). Zadoks & Waibel (2000) classified the various crop protection methods schematically.

Agricultural threats

World-wide a large proportion of the produced yield is lost either by the use of diseased propagules (such as virus-infected seed tubers of potato), thus lowering the vigour of the crop, by competition for external resources (weeds) or internal ones (endophytes, parasitic weeds), by infection of root systems, stem parts, or foliage (thus reducing the yield potential of the crop), or by disease directly affecting harvestable organs. At current use of crop protection measures, including pesticides, about 35% of the agricultural produce is lost before harvest (12% from insects and mite pests, 12% from plant pathogens, 10% from weeds and 1% from mammals and birds). For some major crops this figure can even be substantially higher (e.g. rice: 47% and sugar cane: 54%). After harvest, on average additional losses of 10–15% occur, but in certain instances this figure can be much higher (van Roermund 1997).

Diseases, pests and weeds can either be seed-borne (fungi, nematodes, bacteria), soil-borne (soil fungi, nematodes, weeds), airborne (fungi, insects, weeds) or even water-borne (e.g. bacteria in irrigation water). They have to be controlled. Biotic stresses reduce not only the yield but also the quality of the crop. Especially for export products (ornamentals, seed tubers) it is often mandatory that crops be absolutely free of any organism, or at least the harmful ones.

Weeds have been known throughout agricultural history as a major threat to production, not only for the current crop but also for future crops. Weed control (together with tillage) was the most laborious task of an arable farmer. Diseases were considered as an act of God and pests were simply uncontrollable. When humans discovered that diseases in crops were actually caused by pathogens and that these –

as well as pests and weeds – could be prevented or controlled, crop protection became a science with very clear objectives. Agronomists discovered that there were cures for these problems. It became feasible to develop chemotherapy. In the 1960s, it was clearly stated by leading agronomists that with the clever use of the chemical toolbox almost every agronomic problem could be solved, from pest and weed even to poor soil structure (see, e.g., Kommedahl 1981). This view had its blessings for food production, but gradually resulted in an ecological disruption of agro-ecosystems, that had been more or less stable when people had less control over their environment (Zadoks & Waibel 2000). It took many years before it was realised that the hazardous effects of the use of chemical pesticides would act like a boomerang: there were fewer problems but the remaining ones were extremely difficult to control (see, e.g., Struik & Scholte 1992). Colorado potato beetles became immune to almost all insecticides used in potato cultivation in Long Island (USA) (E.E. Ewing, personal communication), weed populations arose that were resistant against triazines and other herbicides (Ammon *et al.* 1990), even cross-resistance appeared (Kremer 1998), and at the same time the chemicals killed natural antagonists suppressing the pathogens or caused selection for more virulent types (Struik & Scholte 1992; Struik & Bonciarelli 1997). Farmers were actually creating new problems and what was first a solution now became a threat for the sustainability of the natural resources of the farmer. On top of that, the use of chemical crop protectants proved to have a negative impact on the environment outside the agro-ecosystem, on vital resources (biodiversity, drinking water) and on life, especially of organisms at the end of the food chain, including humans (Zadoks & Waibel 2000).

Long-term agronomic prospects of use of synthetic pesticides

Currently there is a general awareness among scientists, farmers, governments, and the public at large that the chemical era of crop protection, despite its success story, has to come to an end and that an ecological approach is more sustainable. Zadoks & Waibel (2000) analysed the history of the use of synthetic pesticides and identified five lessons that can be learned from that history:

(1) High pesticide usage is counter-productive. First of all, it causes uncontrolled external effects on the quality of the environment, in both the short term (e.g. pollution of drinking water) and the long term (e.g. carcinogenic or teratogenic compounds in the environment). Secondly, it destabilises the cropping system, thus eliminating part of the self-regulating power of the system (Almekinders *et al.* 1995). The best described example of the latter is the negative impact of high pesticide usage on rice production, caused by the negative effects on non-target organisms that served as biological control agents (Kenmore 1996). Moreover, continuous use of biocides may trigger resistance to active ingredients in the target organisms (Ammon *et al.* 1990), cross-resistance to other active ingredients (Kremer 1998), the build-up of more destructive populations of target organisms (Fry & Smart 1999) or other pests (Struik & Bonciarelli 1997), or the

more rapid breakdown of the product in the environment (Struik & Scholte 1992).

(2) Synthetic pesticide use requires intensive regulation to limit the external effects. More intense regulation usually results in stricter requirements for registration, resulting in higher costs for R&D and for use, resulting in lower volumes and in fewer options for control of agronomic problems. For an overview of possible policy instruments see Oskam *et al.* (1998).

(3) Since external effects were not initially taken into account, the socio-economic benefits of pesticides (in terms of productivity increases) were initially over-estimated. At first, the standard benefit:cost ratio was estimated to be 4:1 (Headley 1968), but recent estimates indicate that benefit:cost ratios are closer to 1.5:1 (Waibel & Fleischer 1998) or in some cases even smaller than 1:1 (Rola & Pingali 1993).

(4) Early estimations of net agronomic benefits of pesticide use were over-optimistic. They were based on crop loss assessments and older economic studies that had serious methodological flaws. This was partly caused by the pivotal role that the pesticides industry played in creating the pesticide-friendly information environment (Tombs 1993, cited by Zadoks & Waibel 2000).

(5) Intensive use of pesticides made farmers very dependent on them, so that they lost important alternative options. Farmers actually facilitated the reduction of the self-regulating mechanism in the cropping system, making them more dependent on the use of synthetic pesticides, at the same time increasing the profitability of pesticide use (Zadoks & Waibel 2000). Alternatives were no longer used, or even no longer recognised or known. The emergence of very narrow crop rotations (e.g. one potato every two growing seasons) in the peat colonies of The Netherlands is a good example.

This historic analysis suggests that with a similar research input in non-chemical approaches control strategies may be feasible that are agronomically and environmentally more sound, economically competitive and more socially sustainable.

A change in attitude

All over Europe, governments have made plans drastically to reduce the dependence on chemical inputs in agriculture as well as the amount of chemical inputs used (see, e.g., Oskam *et al.* 1998). Biological control has already become an accepted technology in protected cultivation, and, moreover, a symbol of good horticultural practice.

The situation is more complex in field crop production. Typically in arable agro-ecosystems, there is a more or less fixed pattern in the temporal distribution of the crops (Struik & Scholte 1992). Farmers try to maximise the frequency of the most profitable crop or crops. Since a certain rotation is selected on the basis of economy, often crop rotations within a certain region are relatively similar, so that the spatial diversity in crops is also limited. With this lack of diversity in time and space, soil-borne disease problems can easily become severe, and airborne diseases can spread readily as soon as an infection source has been established (Oskam *et al.* 1998).

In arable farming, human control of the environment is much less than in horticulture, and consequently exposure to biotic stresses is more intense and diverse. Controlling one biotic stress may enhance another one. New problems appear with frightening frequency (e.g. bacterial wilt: Elphinstone 1996), or old problems become intractable (e.g. late blight in potato: Fry & Smart 1999). The labour input required to protect the crop without chemical input is enormous (especially in relation to weed control), the crop and financial losses under sub-optimal protection are large and the margins extremely small. Nevertheless, 'biological' farming is becoming more popular and is increasingly promoted by governments because they see it as the best route towards sustainability. Moreover, some farmers experienced in biological or organic farming are doing so well commercially that the potential of the approach has clearly been demonstrated. These aspects provide an extra incentive to reconsider current crop production strategies. We need to develop another approach to the threats and another technology to cope with them.

Agricultural solutions

First and foremost it is necessary to develop a prevention system based on agro-ecological insight. This means that agriculture should seek an ecological approach that makes better use of the buffering and stabilising capacity of the agro-ecosystem itself, in order to help suppress biotic stresses. Vereijken (1997) developed a methodological way of prototyping arable farming systems in interaction with pilot farms that may result in the design of sustainable cropping systems with a strongly reduced use of pesticides.

A logical second approach to the reduction of agronomically and environmentally negative side effects of chemical control is to refine their use by (a) developing better chemicals of which less is needed for effective control and which are less damaging to the environment, and (b) applying them in a technologically more advanced way.

This last aspect of improving application techniques may include making use of precision techniques or using tools to increase the efficacy. A good example is the use of the minimum lethal herbicide dose based on high-tech measurements with chlorophyll fluorescence shortly after application, thus predicting the efficacy of the application. After this system had been tested for two years, most farmers were quite satisfied with the technique and had enough trust in the method actually to apply it (L.A.P. Lotz, Plant Research International, personal communication). This technique can result in a tremendously reduced input of chemicals, but not in a reduction of the dependence of the farm activities on chemical inputs. Although this technological approach of rationalising and refining input of chemicals is an important strategy that should not be ignored or underestimated, it cannot be the only one.

An integrated approach makes use of a wide range of preventive or non-chemical control measures, trying to avoid chemical input and using biocides only where the other tools fail. Agricultural science is working on developing a tool kit that will go a long way in this direction, and will make it increasingly rare for chemicals to be needed. Unpublished research of the former Department of Agronomy of Wageningen University, for example, showed that in a cropping system with a very high frequency of potato cultivation and an intended heavy pressure of a variety of potato pests and diseases, a diversity of non-chemical control measures could make the cropping system healthy again and restore the yield to the level of the treatment with full chemical control. Research is currently also working on risk assessments related to restricted use of chemicals in the conversion to more sustainable arable farming (e.g. de Buck 2001).

The most extreme approach is when a farmer decides to refrain from any use of biocides and accepts the incidental failure of the crop or the incidental high yield and quality losses.

All such approaches require a systems approach in which the farmer makes use of ecological principles at the systems level to prevent, suppress and control biotic stresses. This is partly a return to old principles, but now using new technologies. In no respect can it be considered as a return to old technology. In the past, farmers relied on and actively enhanced the stability of the agro-ecosystem, because they could not control the fluctuations that would otherwise occur. The difference now is that we are currently seeking stability at a much higher level of production, and thus a much larger intensity of use and a higher rate of turnover of resources. We do so on the basis of much more scientific knowledge, while using much more advanced equipment and tools.

A new approach

The use of advanced ecological knowledge by agronomists is fairly recent, especially at higher levels of the hierarchy of the agro-ecosystem. Ecologically sustainable systems of management of pests, diseases and weeds should include three elements: prevention, decision making and control. These three elements will be discussed in detail.

Prevention

An essential element of proper management of diseases, pests and weeds is prevention. It is crucial to design land use systems, farming systems and cropping systems in such a way that problems will hardly occur.

Land use systems can be re-designed in such a way that the spatial and temporal distribution of crops of the same species is less conducive to the spread of the disease or pest (Oskam *et al.* 1998). Regional distribution of crops can be regulated, e.g. minimum distances between crops of the same species, isolation of seed potato producing areas from aphid populations, separation of propagule production from ware production (Struik & Wiersema 1999; Caldiz *et al.* 2002). In this way the spread of airborne diseases can be limited. Crop rotations – a crop rotation is a more or less fixed pattern in the succession of crops on a certain field – can be widened (Struik &

Bonciarelli 1997). The spatio-temporal consequences of these two measures on crop-ping patterns, and thus occurrence of biotic stress, can be optimised (Oskam *et al.* 1998).

Farming systems should be designed in such a way that the chances of re-infection or spread are minimal. This may include the design of farms in such a way that fields of easily infected crops are well surrounded by ecological safe-havens for antagonists, natural enemies and other beneficial organisms (Smeding 2001). Crop rotation should take into account the multiplication of soil-borne pathogens or should allow optimal control of weed populations (Struik & Bonciarelli 1997). Important elements of a crop rotation include: which crops are grown in a rotation, the frequency with which each crop is grown, and in what sequence crops are grown (Struik & Scholte 1992; Struik & Bonciarelli 1997). Although it is difficult to set out general rules for a good crop rotation, one should at least consider the effect of preceding crops on physical, chemical and biological soil fertility, the sensitivity of following crops to these effects and the accumulation of effects over cropping systems (Struik & Bonciarelli 1997).

Cropping systems can be diversified by using strip cropping, mixed cropping, vari-etal mixtures, enhancing associated biodiversity in the system, or even including and maintaining natural, disease or pest suppressing elements in the farming system (Almekinders *et al.* 1995; Smeding 2001). For example, inter- and relay cropping of wheat and cotton in China, to advance a dense population of the seven-spot beetle controlling aphids damaging the cotton crop, has proven successful and has become a common prevention strategy in north China (Xia 1997).

Prevention can be optimised by maximisation of the use of natural processes in the cropping system, suppressing the harmful organisms, including the development of antagonists, optimising the diversity of the system, and stimulating the recycling of internal resources. Instruments and tools to achieve that are rotation-specific, but may include:

(1) Farm hygiene. It has been known for a long time that in newly reclaimed land, some farmers managed to stay free from harmful nematodes much longer than did surrounding farmers (Kuiper 1977). Apparently they had developed better strategies to maintain the hygiene on the farm. An important element of farm hygiene is the use of clean seed or planting material and maintaining temporal and spatial separation between crops of the same species (control of volunteers!) (Struik & Wiersema 1999).

(2) The synergistic and antagonistic effects that occur in a cropping system. Ecolog-ical approaches to crop protections are viable, as has been proven in an experiment with a 50% frequency of potato in the presence of many different soil-borne pests and diseases. Despite the high frequency of potato the diseases and pests could all be suppressed by a perfectly designed system of non-chemical preventive methods, including the cultivation of catch crops, amendments of organic matter to enhance populations of antagonists, and so on. Other exam-ples are described by Struik & Bonciarelli (1997).

(3) The design of cultural practices that support such ecological processes, such as delayed planting to reduce weed growth or even prevent their seed set (e.g.

Keeley & Thullen 1983), the removal of crop residues or plant debris (Mol 1995), management of soil organic matter (Lootsma 1997), and soil tillage strategies (e.g. Yenish *et al.* 1996).

(4) The optimisation of other inputs so that a crop can be created with a functional condition that will assist it in withstanding attacks of pathogens or that will increase the damage threshold, e.g. providing extra nitrogen to a wheat crop to reduce or to overcome the attack by the take-all syndrome (Struik & Bonciarelli 1997).

(5) Breeding for tolerance, e.g. by selecting for specific sugarbeet plant types that are more competitive against weeds, and (polygenic) durable resistance, e.g. against blights.

If, despite proper prevention methods, problems still do occur it is crucial to make a wise decision on crop protection. That brings us to the second element: the improvement of decision making.

Decision making

There are three major types of decision making in crop protection:

(1) strategic decisions, in which long-term developments are taken into account
(2) tactical decisions, in which decisions are made that affect the management of a crop over an entire growing season
(3) operational decisions in the field, in which it is decided what to do when and how.

Until recently the emphasis in agriculture was very strongly on the last type of decision making (Zadoks & Waibel 2000; see also above). Based on observations the occurrence of a problem was established and a solution was found by selecting the right chemical. In some cases, when a problem was likely to appear, even preventive pesticide sprayings took place, long before any large-scale infestation of the pests was observed. In some cases (e.g. in pre-emergence weed control or in late blight control in potato) this could even prove to be an efficient strategy, also in terms of use of the active ingredient.

Operational decision making has changed with the introduction of the damage threshold concept (Pedigo *et al.* 1986; Van Roermund 1997). Only in cases when the damage reaches, or is expected to reach, a certain economic limit, application of a pesticide is advisable. Operational decisions, however, should include not only the immediate effects on the current crop, but also the mid-term and long-term effects. This is the case, for example, in weed control, where not only must the immediate competition between crop and weed plant be shifted in favour of the growth of the crop plant, but also the production of the week's survival structures – seeds, spores, tubers – must be taken into account, since they form a threat to future crops (Struik & Bonciarelli 1997). In this way the operational decision becomes a tactical or even a strategic one. Decision making in weed control is in fact long-term population management. Therefore, the threshold concept does not hold for weed management and should be restricted to pests and diseases, which have to colonise the agro-ecosystem every crop cycle.

The decision making process for tactical and operational decisions in management of pests, diseases and weeds can be represented as illustrated in Fig. 2.1. It involves strategies to determine if, when, where and how control should take place. The first assessment required for decision making is the seriousness of the infestation. On the basis of this assessment, predictions must be made on the yield effect of the infestation in the current crop and its potential after-effects in future crops. The infestation also defines the cultivation techniques and possible control technologies that may reduce its negative effects. On the basis of the prediction of the expected yield loss, the risks of future yield losses, and the instruments available to the farmer to control the pest or disease or to reduce its effects, a decision needs to be made on whether control is needed at all, and if so, when action should take place and what technique(s) should be used. This decision can only be taken properly if both the direct costs of control and the external effects (e.g. impact on the environment) are taken into account ('Criteria' in Fig. 2.1). In any case, the decision will have an impact on the development of the infestation level.

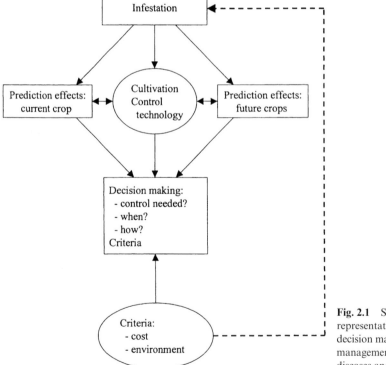

Fig. 2.1 Schematic representation of the decision making process in management of pests, diseases and weeds.

To allow *rational* decision making, the severity of the infestation and its effects must be known quantitatively, e.g. on the basis of a decision support system. For soil-borne diseases this quantitative information must be available before the growing season, for seed-borne pathogens at planting or sowing, and for airborne or water-borne pathogens early during the growing season. Knowledge of the severity of the infestation can help to predict crop yield losses (both in terms of quantity and quality), but also the multiplication of the harmful organisms (weed seed, nematodes, etc.). Criteria must be defined, based on objectives, planning and risk attitude of the farmer, to allow economic decisions to be made. At the same time, costs for control can be assessed quantitatively and the long-term damage of taking no control measures can be estimated. Efficacy of control methods in relation to the technology and the timing can also be determined, as well as the possible side effects of the application.

Important questions are:

- Is control needed?
- When is control needed?
- Where should control take place (in patches or throughout the field)?
- How should control take place?

To answer the question 'Is control needed?', threshold densities for control can be taken into consideration, although it is not always easy in all cases (see earlier). A threshold level can just be tolerated, based on economic criteria or risk assessment. There are different concepts for thresholds for operational, tactical and strategic decisions (see, e.g., de Buck 2001). The differences are associated with the time horizon of the occurrence and duration of the biotic stress.

Timing of control is crucial, but it also depends on the techniques of control. Especially in mechanical control of weeds, timing is of the essence. In weed control the concept of minimum lethal herbicide dose (MLHD) is gaining some popularity. The MLHD depends on weed development and therefore, if chemical control is needed, it is better to do it early, since an early application will save active ingredients and costs.

Control can be local – a row application of a herbicide in association with mechanical control between rows, control of nematodes in patches – but local control requires good diagnostic tools and close observation, especially in the case of soil-borne problems.

From the point of view of the environment it is best to try to control problems with biological means first, then with mechanical means and finally – if other methods fail – with chemical means. However, if a poor control by non-chemical means results in a large problem requiring a large input of chemicals to control it, preventive use of chemicals may be wise. Future options here may lie in the development of *self-learning systems*, which develop *site-specific knowledge* and thus assist in acquiring *site-specific management options* (*cf.* Swinton in this volume).

Control techniques are rapidly developing; precision management is not far away. Optical techniques for weed control and control of pests and diseases that cause clearly visible above-ground plant symptoms may be within reach in the near future.

Control

Control of pests, diseases or weeds may occur through different methods: mechanical, chemical, biological (through a one-to-one relation between pathogen and antagonist), ecological (involving a much wider range of harmful and beneficial organisms) or integrated (using different methods simultaneously and in good harmony). Whatever method is selected, control should take place with precision, with a high efficacy and preferably with as few chemicals as possible.

In Table 2.1, we have summarised the main categories of biotic stress agents and what the major options are for prevention, decision making and control. This table shows that there is a large difference among groups of agents and that in some cases there is still a long road ahead of us. For more details see below.

Table 2.1   Overview of main elements of non-chemical options in the prevention, decision making and control of different categories of causal agents of biotic stress. ML(H)D – minimum lethal (herbicide) dose.

	Prevention	Decision making	Control
Fungi	Clean seed Resistance Separation Cultural practice Crop rotation Hygiene Quarantine	Damage threshold Detection Warning systems	Precision (MLD) Selective means Integrated control Antagonists Fungicides still dominant
Nematodes	Crop rotation Resistance Clean seed Hygiene Quarantine	Quantitative detection Damage threshold	Trap crops Biological control Luring compounds Physical methods Organic matter and antagonists
Viruses and bacteria	Crop rotation Quarantine Roguing Clean seed Buffer systems Hygiene Irrigation control	Date of haulm killing	Vector control Roguing Haulm killing
Weeds	Crop rotation Cultural practice Weed suppression Hygiene	Strategic options Precision	ML(H)D Application techniques
Insects	Buffering Clean seed Hygiene	Thresholds	Ecological techniques Enemies

Agronomic instruments to reduce use of pesticides

As stated above, crops can be attacked by organisms that are either seed-borne, soil-borne, airborne or water-borne. If chemical control is necessary these different types of organisms will require different types of crop protectants, and their formulation, method and frequency of application, amount of crop protectant to be applied, and their impact on the environment will also differ (Oskam *et al.* 1998).

Strategies and instruments designed to reduce the amount applied can focus either at the causal organism, the crop or the technological aspects of the actual use. Because we have already partly discussed the agronomic options for reductions in use in previous sections (see also earlier sections for relevant references), we will here summarise the options.

With regard to the causal organisms

A first option is to avoid the problem, or the spread of the problem, by control at regional level (e.g. by spatially separating the production of transplants and the market-oriented food production of leek often affected by airborne fungi), by farm hygiene (control of nematodes), control outside the cropping systems (potato heaps and late blight), or adjustments within the cropping system (improving habitats for antagonists). It is also possible to take measures to reduce the density of survival structures of biotic stresses (e.g. by removing plant debris that serves as a source of inoculum), or measures to reduce their ability to attack the crops (isolation, adjusting crop cycle or sowing time, haulm killing). Monitoring of epidemics and perfecting the damage threshold concept can support decision making so as to prevent severe outbursts.

With respect to the crop

The chances of a crop or plant parts being attacked can be reduced by mechanisms including cultivar rotation, mulches, intercropping, creating closed canopies at an early stage, irrigation, growing repellent plants or trap crops or placing traps, and so on. The success of infection can be reduced by changing the sensitivity of the crop (e.g. by changing either the crop structure, the microclimate within the crop, or the chemical composition of the crop). It is often productive to make the crop plants less vulnerable to the attack or make a quick recovery possible, e.g. by optimal fertilisation.

With respect to the use of crop protectants

Based on monitoring, warning systems, weather and yield predictions, decision support systems, and risk and damage assessments, the use of crop protectants can be limited to those conditions in which a strong positive effect of use can be expected. Targeted sprays (based on detection and precision agriculture techniques) can help further to reduce actual quantities used. Use may also be optimised by better labelling, increased efficacy, improved application techniques and other measures reducing risks for food safety, emission, environmental hazard and other damaging effects.

Policy instruments to enhance the reduction of pesticides

To link this agronomic view to the views of more policy oriented authors in this book, we have translated these agronomic instruments into apolitical policy instruments, without taking into account whether or not it would be practical to enforce these instruments. Policy instruments can be seen as limitations and restrictions, and as stimulating technological change, supported by research (see also Oskam *et al.* 1998).

Policy instruments in the first category could include:

- providing chemicals on recipe only;
- forbidding their use in certain cases;
- restricting their use to certain crops or cultivars;
- proper labelling and mandatory rotation of pesticides for frequently and persistently occurring problems;
- enforcing strict rules on farm hygiene, cropping frequency and cropping sequence;
- control of the spread of cultivars and stimulation of cultivar rotation or cultivar mixtures;
- improved control on the health status and distribution of propagules;
- development of agro-technological indicators (yardsticks) for progress in new cropping technologies.

Policy instruments in the second category may include:

(1) Changing pesticide technology by enhancing a proper balance between preventive and curative pesticides, quality guarantee systems of pesticides, development and subsidising of alternative application techniques, development of techniques to reduce harmful effects of pesticides, and development of better techniques to allow application of chemicals in patches, including specific site management strategies and precision agriculture.

(2) Changing agronomic practices by enhancing the use of green manure crops, repellent crops, lure crops, trap crops and killing crops with a beneficial effect on soil life, use of mulches, development and use of techniques to enhance possible decline effects in crop rotations, re-thinking the possibilities of re-introducing intercropping in Western agriculture, and enhancing use of proper crop management to reduce disease spread and increase tolerance, resistance or recovery of a crop.

(3) Changing attitudes to pesticide use by stimulating mechanical control measures, development and use of warning systems, development, maintenance and use of decision support systems, and by improving existing systems of threshold levels of diseases and detection and selection techniques.

Closing remarks

The chemical era of crop protection will come to an end, at least for certain pathogens. Our ecological and agronomic knowledge on how to prevent and control pests, diseases and weeds has increased. Novel techniques are under development that will improve the resistance and tolerance of crops, will increase efficacy of control and will help to predict the chances of yield loss and assist in decision making on control measures. Nevertheless, more can be done to upgrade our ecological understanding at higher levels of aggregation and on transferring this knowledge to the individual grower.

References

Almekinders, C.J.M., Struik, P.C. & Fresco, L.O. (1995) The need to study and manage variation in agro-ecosystems. *Netherlands Journal of Agricultural Science*, 43, 127–42.

Ammon, H.U., Struik, P.C. & Stamp, P. (1990) Silomaisproduktion in klimatischen Grenzlagen. VII Agrarökologische Aspekte des Anbaus. *Kali-Briefe (Büntehof)*, 20, 293–309.

de Buck, A. (2001) *The role of production risks in the conversion to more sustainable arable farming.* Doctorate thesis, Wageningen University.

Caldiz, D.O., Haverkort, A.J. & Struik, P.C. (2002) Analysis of a complex crop production system in interdependent agro-ecological zones: a methodological approach for potatoes in Argentina. *Agricultural Systems*, 73, 297–311.

Elphinstone, J.G. (1996) Survival and possibilities of extinction of *Pseudomonas solanacearum* (Smith) in cool climates. *Potato Research*, 39, 403–10.

Fry, W.E. & Smart, C.D. (1999) The return of *Phytophthora infestans*, a potato pathogen that just won't quit. *Potato Research*, 42, 279–82.

Headley, J.C. (1968) Estimating the productivity of agricultural pesticides. *American Journal of Agricultural Economics*, 50, 13–23.

Keeley, P.E. & Thullen, R.J. (1983) Influence of planting date on the growth of black nightshade (*Solanum nigrum*). *Weed Science*, 31, 180–84.

Kenmore, P. (1996) Integrated pest management in rice. In: *Biotechnology and integrated pest management* (ed. G.J. Persley), pp. 76–97. CABI Publishing, Wallingford.

Kommedahl, T. (1981) *Proceedings of Symposia, IXth International Congress of Plant Protection,* Washington, DC., USA, 5–11 August 1979. Burgess Publishing, Minneapolis.

Kremer, E. (1998) *Fitness of triazine susceptible and resistant* Solanum nigrum *L. in maize.* Doctorate thesis, Wageningen Agricultural University.

Kuiper, K. (1977) *Introductie en vestiging van plantparasitaire aaltjes in nieuwe polders, in het bijzonder van* Trichodorus teres. *Mededelingen Landbouwhogeschool*, 77, 4. PhD thesis, Wageningen University.

Lootsma, M. (1997) *Control of Rhizoctonia stem and stolon canker of potato by harvest methods and enhancing mycophagous soil mesofauna.* Doctorate thesis, Wageningen Agricultural University.

Mol, L. (1995) *Agronomic studies on the population dynamics of* Verticillium dahliae. Doctorate thesis, Wageningen Agricultural University.

Neumann, R. (1997) Chemical crop protection research and development in Europe. In: *Perspectives for agronomy. Adopting ecological principles and managing resource use.* (eds M.K. van Ittersum & S.C. van de Geijn), pp. 49–55. Elsevier Science, Amsterdam.

Oskam, A.J., Vijftigschild, R.A.N. & Graveland, C. (1998) *Additional EU Policy Instruments for Plant Protection Products.* Wageningen Pers, Wageningen.

Pedigo, L.P., Hutchin, S.H. & Higley, L.G. (1986) Economic injury levels in theory and practice. *Annual Review of Entomology*, 31, 341–68.

Porceddu, E. & Rabbinge, R. (1997) Role of research and education in the development of agriculture in Europe. *European Journal of Agronomy*, 7, 1–13.

Rola, A. & Pingali, P. (1993) *Pesticides, rice productivity, and farmers' health. An economic assessment.* IRRI, Los Baños.

Smeding, F.W. (2001) *Steps towards food web management on farms.* Doctorate thesis, Wageningen University.

Sonnewald, U. & Herbers, K. (2001) Plant biotechnology: Methods, goals and achievements. In: *Crop science: Progress and prospects* (eds J. Nösberger, H.H. Geiger & P.C. Struik), pp. 329–50. CABI Publishing, Wallington.

Struik, P.C. & Bonciarelli, F. (1997) Resource use at the cropping system level. *European Journal of Agronomy*, 7, 133–43.

Struik, P.C. & Scholte, K. (1992) Crop rotation. In: *Proceedings Congress Agriculture and Environment in Eastern Europe and The Netherlands* (ed. J.L. Meulenbroek), pp. 395–423. Wageningen University, Wageningen.

Struik, P.C. & Wiersema, S.G. (1999) *Seed potato technology.* Wageningen Pers, Wageningen.

Tombs, S. (1993) The chemical industry and environmental issues. In: *Business and environment: Implications of the new environmentalism* (ed. D. Smith), pp. 131–49. Chapman, London.

Van Roermund, H.J.W. (1997) Pests and diseases. In: *Agro-ecology* (eds D. Kleijn, C.A. Langeveld, M.J. Kropff & W. Joenje), Chapter 9. Department of Plant Sciences, Wageningen University, Wageningen.

Vereijken, P. (1997) A methodological way of prototyping integrated and ecological arable farming systems (I/EAFS) in interaction with pilot farms. *European Journal of Agronomy*, 7, 235–50.

Waibel, H. & Fleischer, G. (1998) *Kosten und Nutzen des chemischen Pflanzenschutzes in der Deutschen Landwirtschaft aus gesamtwirtschaftlicher Sicht.* Vauk Verlag, Kiel.

de Wit, C.T. (1992) Resource use efficiency in agriculture. *Agricultural Systems*, 40, 125–51.

de Wit, C.T., Huisman, H. & Rabbinge, R. (1987) Agriculture and its environment: Are there other ways? *Agricultural Systems*, 23, 211–36.

Xia, J.Y. (1997) *Biological control of cotton aphid* (Aphius gossypii *Glover) in cotton (inter)-cropping systems in China; a simulation study.* Doctorate thesis, Wageningen Agricultural University.

Yenish, J.P., Fry, T.A., Durgan, B.R. & Wyse, D.L. (1996) Tillage effects on seed distribution and common milkweed (*Asclepias syriaca*) establishment. *Weed Science*, 44, 815–20.

Zadoks, J.C. & Waibel, H. (2000) From pesticides to genetically modified plants: History, economics and politics. *Netherlands Journal of Agricultural Science*, 48, 125–49.

Chapter 3
The Authorisation of Pesticides in the Light of Sustainability

Elizabeth Vogelezang-Stoute

Introduction

A pesticide product has to be authorised before it is allowed onto the market in a member state of the European Union. In the authorisation process, the efficacy and the possible adverse effects of the use of the pesticide are evaluated. Authorisation can be seen as a legal instrument whose purpose it is to test a pesticide before it is brought onto the market, in order to ensure that it meets certain requirements. Authorisation is given for a limited period only, so that afterwards there will be an opportunity to conduct a review according to current scientific standards.

Although pesticides have to be authorised before being marketed and used, this does not seem to be preventing the use of pesticides from having adverse effects on the environment. Therefore the question arises whether inadequacies in the authorisation instrument or in the application of this instrument contribute to these adverse effects. What bottlenecks can be distinguished, and can options be considered for conducting the authorisation process in a way that could contribute to a sustainable use of pesticides? The purpose of this chapter is to identify some of the bottlenecks that prevent adequate environmental protection and to consider possible solutions for achieving a more sustainable use of agricultural pesticides. An analysis will be given of the authorisation process as laid down in Directive 91/414/EEC on the marketing and use of plant protection products (henceforward referred to as 'Directive').[1] The authorisation practice will be illustrated for the EC level and for the national (i.e. Dutch) level. To illustrate other approaches to pesticide regulation reference will also be made to the US federal registration system and to a Swedish review project.

Certain aspects of the authorisation process under the Dutch Pesticides Act will be described to illustrate some of the changes introduced at the national level as a result of the Directive. An EU member state has to comply with the Directive by transposing it into the national legislation. Only to the extent allowed by the Directive and the EC Treaty will a member state have discretion to put other national measures on this subject into place.

Although one could wonder whether pesticide use can ever be considered in terms of sustainability, my assumption is that pesticide use is 'sustainable' not only when the product meets certain requirements but also when the application is limited to a strict minimum and priority is given to methods that are the least harmful from an environmental point of view. This interpretation of sustainable use refers to the definition of 'integrated control' in article 2 of Directive 91/414/EEC.

The second section roughly sketches the historical background behind pesticide legislation. The third section describes the structure and some main elements of Directive 91/414. The implementation of the Directive in the Dutch Pesticides Act is the subject of the fourth section. Then the positive contributions that the authorisation system can possibly bring to a more sustainable use of pesticides will be described. The sixth section identifies some of the bottlenecks in the process, with conclusions and some recent policy developments presented in the final section.[2]

Historical background and scope

Legislation on the authorisation of pesticides in many states dates from around the middle of the last century. In The Netherlands the 1947 act on pesticides and fertilisers focused mainly on the efficacy of the pesticide products.[3] The current Dutch Pesticides Act (Act) dates from 1962.[4] During the course of subsequent amendments the focus of the Act shifted. At first, it was consumers and occupational health that needed to be protected; later, the environment itself was seen as in need of protection. Protection of the environment became an explicit goal of the Act in the beginning of the 1970s. By that time some effects of long-term use had become clear, such as bio-accumulation of certain persistent substances in birds at the end of the food chain.[5]

In the United States the registration of pesticides is based on the 1947 Federal Insecticide, Fungicide and Rodenticide Act (FIFRA).[6] The 1910 Insecticide Act was the first US statute to govern pesticide use. It contained standards for the purity of substances, but did not require product registration (Formica & Miller 1999). However, it took a very long time – until the beginning of the 1970s – before the federal registration became more than a rather empty formality. One of the problems of the 1947 Act was that the Secretary of Agriculture could not refuse registration, another that the FIFRA only concerned interstate commerce. In 1972 the scope of the federal law was extended. The Federal Environmental Pesticides Control Act of 1972 changed FIFRA into the act which is the basis of today's FIFRA (Miller 1997). In the meantime the adverse effects of using pesticides had arisen as a topic on the American public agenda. Important steps to public awareness included the publication of *Silent Spring* by Rachel Carson in 1962, and lawsuits by environmental groups, which in the 1970s contributed to the suspension or cancellation of the registration of several major pesticides (Miller 1997; Formica & Miller 1999). Before 1972 registrations were handled by the United States Department of Agriculture, but the 1972 amendment made environmental protection, including human health, the main goal of FIFRA registrations by the Environmental Protection Agency. The 1996 Food Quality Protection Act (FQPA) resulted in another important amendment to FIFRA registration requirements. The FQPA strengthened the requirements regarding food residues, including those resulting from the cumulative effects of pesticides.[7] The FQPA amended both the FIFRA and the Federal Food Drug and Cosmetics Act (FFDCA). The residue standards set by the FFDCA have to be applied under the FIFRA.

The pesticide legislation of various European Union member states changed considerably as a result of the 1991 Directive on the marketing and use of plant protection products. The Directive harmonised the procedures and criteria for making decisions on plant protection products at the national level. At the European Community (EC) level it also created a system for authorising the active substances in these pesticide products.

Directive 91/414 concerns the authorisation of *agricultural* pesticides, in EC terminology, plant protection products. Non-agricultural pesticides, biocidal products, are regulated by Directive 98/8/EC, on the marketing and use of biocides (see Cardonnel & van Maldegem 1998a; 1998b; Vogelezang-Stoute 1999).[8] Some of the differences between these two Directives will be discussed. These two directives are central to the marketing and use of pesticides. There are, however, many other EC directives and regulations that, although outside the scope of this analysis, have an influence on the marketing and use of pesticides, including directives and regulations on residues, on classification, packaging and labelling, on certain active substances and on agri-environmental measures.[9]

Directive 91/414/EEC

Objectives and structure

The aim of Directive 91/414 is to remove trade barriers by harmonising the legislation of member states on plant protection products in a way that ensures a high standard of protection against adverse effects of pesticide use. The Directive's preamble states that plant protection products should not be put on the market unless officially authorised. The authorisation requirements must guarantee a high standard of protection to prevent the authorisation of products whose risks to health, groundwater and the environment have not been appropriately investigated. Moreover, ninth recital of the preamble to the Directive assigns higher priority to the protection of human and animal health and the environment than to the objective of improving plant production.

The Directive creates a dual authorisation system. The authorisation of active substances takes place at the EC level; pesticide products are authorised at the national level. An active *substance* is the ingredient that gives a product its pesticidal effect. A pesticide *product* is the form in which a pesticide is put onto the market. In a pesticide product an active substance is usually combined with other substances (e.g. solvents) to give a product its particular form. Many different products may be derived from one active substance; each of these products requires authorisation. During the implementation of the Directive around 800 active substances of agricultural pesticides were on the market in the European Union. The two authorisation processes are linked by the requirement that the EC authorisation of the active *substance* is a condition for the national authorisation of the pesticide *product*. An active substance is authorised when it is included (i.e. listed) in Annex I of the Directive (article 5). The Directive also provides for the criteria and procedures for the authorisation of products at the national level (articles 4 and 9).

Furthermore the Directive creates a mutual recognition system (article 10). This means that, in principle, a product authorised by one member state will, if applied for, also be authorised by another member state. The other member state can only decide otherwise if the relevant agricultural, plant health or environmental conditions between two member states are not comparable. A condition for mutual recognition is that the active substance is included in Annex I of the Directive. Mutual recognition also applies to decisions on the acceptance of tests and analyses done in other member states. This aspect does not require the inclusion of the active substance in Annex I.

Decision making on active substances

The authorisation procedure for an active substance as laid down in the Directive (articles 6, 19, 20 and 21) may, very briefly, be described as follows. The procedure starts at the member state level, when an applicant submits an application, including the required dossiers, to the competent authority in a member state. The applicant must submit a dossier with the data required in Annex II (for active substances) and a dossier with the data required in Annex III (for pesticide products) for at least one product containing this substance (article 6(2)). The member state ensures that these dossiers satisfy the requirements of the Directive's Annexes and that they are forwarded to the other member states and the European Commission (the 'Commission'). The Commission makes the decision on whether or not to approve the active substance, but the Commission first refers the matter to the EC Standing Committee on Plant Health (the 'Committee'). The Committee is made up of representatives of the member states and is presided over by a Commission representative. The Committee provides its opinion; the Commission then makes its decision. If the Commission's measures are not in accordance with the Committee's opinion, the Commission must submit the matter to the EU Council. If the Council does not make any decision within a certain period of time, the Commission will adopt its proposed measures (article 19). Non-inclusion (i.e. non-listing) of a substance in Annex I implies that a product containing this substance cannot be authorised by a member state. For products already on the market, during the transitional regime, the authorisations will have to be withdrawn. The inclusion (i.e. listing) of a substance in Annex I means that the first condition for product authorisation by a member state has been fulfilled.

In addition to this general procedure for existing substances special review procedures have been laid down in Commission regulations.[10] 'Existing substances' are the active substances of plant protection products that were already on the market two years after the date of notification of the Directive (article 8(2)), i.e. after 26 July 1993. The starting point is a notification by the producer who is seeking to have an active substance included in Annex I. For each substance for which notification has been given, a member state is designated as the rapporteur.[11] The 'rapporteur member state' examines and evaluates the dossiers and reports to the Commission. Note that it is the member state that carries out specific scientific and technical tasks that are the basis for the EC decision making. Then a peer review at EC level is conducted by experts from several member states. In the early years it appeared that there were considerable gaps in the reports and that the quality of the initial assessment reports varied considerably (SANCO/2692/2001:7).

The next step is evaluation by the Committee, which has an Evaluation working group and a Legislation working group. Scientific questions are dealt with in the Scientific Committee on Plants. Finally, the Commission–advised by the (Standing) Committee–decides, first, on the completeness of the dossiers and, second, on whether or not to approve the active substance. In practice the pace of this decision making has proven to be very slow. By mid-2001 only 25 active substances (11 existing substances and 14 new substances) had been included in Annex I. For 16 active substances the decision reached was not to include them, in most cases because of incomplete data, in other cases because the substances did not meet the requirements. From these figures one can conclude that it will take some time before the approximately 800 existing active substances that are on the market in the European Union will all be reviewed.

The Directive's twelve-year transition period (ending in 2003) for the review of the existing active substances will not be long enough to conduct the review process for all the substances. According to article 8(2) of the Directive the twelve-year period can be extended for certain substances. In 2000 the Commission concluded that for most of the active substances of the first phase (90 substances) the information submitted had been insufficient. Therefore, time limits were laid down and restrictions were imposed for the submission and acceptance of new studies.[12] In order to speed up the review process the Commission introduced stricter time limits for the next phases of the work programme. The rules laid down in the Regulation for the second and third phase of the work programme contain more strict time limits, especially regarding incomplete dossiers.[13]

The Directive's criteria for the authorisation of an active *substance* are similar in parts to the criteria for the *products*, which are described hereafter. With regard to the substances, the effects of the residues, the acceptable daily intake, the acceptable operator exposure, the distribution in the environment and the impact on non-target species are singled out for particular attention (article 5). Decision making on these aspects has to take place according to current scientific and technical knowledge. Other than for the product authorisations by member states, no specific principles are laid down for the evaluation and decision making for the active substances. The Commission decisions on inclusion or non-inclusion of the active substances give little information on how the review has taken place or on what grounds the decision was made. Member states do have to keep the finalised review reports available for consultation by interested parties.[14]

Decision making on plant protection products

The Directive includes the procedures and criteria that a member state must apply in making a decision on whether to authorise a product. The criteria focus on efficacy, on the adverse effects of use, on analytical methods of determining residues and on physical and chemical properties. Current scientific and technical knowledge must be applied to determine that the products meet these criteria. The adverse effects criteria require that the product, when properly used, not cause:

- an unacceptable effect on plants or plant products;
- unnecessary suffering and pain to vertebrates that are to be controlled;
- a harmful effect on human or animal health, directly or indirectly, or on groundwater;
- an unacceptable influence on the environment, having particular regard to the contamination of water, including drinking water and groundwater, and to its impact on non-target species;
- toxicologically and environmentally significant residues that are not determinable by appropriate methods (article 4(1)).

The repeated reference to groundwater in these criteria seems to be an indication that groundwater is meant to be subject to very strict protection. Use may cause neither a 'harmful' nor an 'unacceptable' effect on groundwater. While the term 'unacceptable' suggests a certain balancing of interests, the word 'harmful' does not appear to contemplate this balancing exercise. For the interpretation of these criteria, however, the Uniform Principles have to be taken into account.

The Uniform Principles, which are laid down in Annex VI of the Directive, contain requirements for the evaluation and decision making by the member states. They aim at ensuring that the member states apply the criteria in an equivalent manner and provide the high level of protection required by the Directive. The Uniform Principles include principles of both general and specific application. Uniform Principle C.1.3. is an example of a *general* principle:

> Member states shall ensure that the authorised amounts, in terms of rates and number of applications, are the minimum necessary to achieve the desired effect even where higher amounts would not result in unacceptable risks to human or animal health or to the environment.

This principle illustrates a preventative approach: even where there is no 'unacceptable risk' the amounts used have to be the minimum necessary. The Uniform Principles also contain *specific* standards for the pesticide and its metabolites (e.g. environmental standards for the persistence in soil, the expected concentration in groundwater and the impact on non-target organisms), as demonstrated by this excerpt from Uniform Principle C.2.5.2.1:

> Where there is a possibility of birds and other non-target terrestrial vertebrates being exposed, no authorisation shall be granted if: the acute and short-term toxicity/exposure ratio for birds and other non-target terrestrial vertebrates is less than 10 on the basis of LD50 or the long-term toxicity/exposure ratio is less than 5, unless it is clearly established through an appropriate risk assessment that under field conditions no unacceptable impact occurs after use of the plant protection product according to the proposed conditions of use, …

While these standards seem rather strict, there are various uncertainties inherent in the Principles, which could undermine the effective application of these standards. One of these uncertainties is the use of 'disclaimers'. An example is given in the last standard described above. The disclaimer implies that the standard will not apply if risk assessment clearly establishes that under field conditions there is no unacceptable impact. The Principles do not provide guidance as to when there is 'no unacceptable impact'.

Another uncertainty is the use of vague descriptions such as 'relevant metabolites' which have to be taken into account, for example for the concentration in groundwater (Uniform Principle C.2.5.1.2.). Of course, the interpretation of what is meant by 'relevant' is crucial for defining this requirement. There are many uncertainties of this kind, where further interpretations are required.

At the EC level guidelines for interpreting standards and other requirements are given in 'guidance documents'. These guidelines are given to applicants for preparing the submission of data and to the member states for the evaluation of dossiers. What a standard really entails in practice may depend on these guidance documents. Therefore it is remarkable that the Directive does not regulate a procedure for how these guidelines are to be set. These documents might be seen as opinions of the Commission services, elaborated upon in cooperation with member states. However there are also guidance documents which 'do not necessarily represent the view of the Commission Services', such as the Guidance Document on Voluntary Mutual Recognition of Minor Use Authorisations (SANCO/2971/2000). A transparent procedure for the development and publication of these documents seems necessary, particularly because the evaluation and decision making as laid down in the Uniform Principles require risk analyses and the use of calculation models. For these, the guidance documents could be essential for producing a proper and comparable outcome, one reached by the application of current scientific and technical knowledge. On this point of guidance documents, Directive 98/8/EC is clearer. Article 33 of this directive requires that 'Technical notes for guidance to facilitate the day-to-day implementation of this Directive' be drawn up in accordance with a Committee procedure.

Discretionary powers left with the member states

One could wonder what discretion Directive 91/414 has left the member states to take their own pesticide measures at the national level. Not only do they have to apply criteria and procedures as described above, they also have to take into account the EC decisions on the active substances. Therefore the effect of the Directive is not only harmonisation, but also includes centralisation. Furthermore, once an active substance is listed in Annex I, the system of mutual recognition, mentioned before, will apply. If a member state and an applicant do not agree on whether the circumstances between member states are comparable, the final decision will be made at the EC level. This means that not much discretion has been left at the national level.

The Directive does give some discretion to a member state, namely to authorise products based on *new* active substances not yet reviewed for Annex I. This derogation from the authorisation criteria aims to stimulate innovation in the field of pesticide products (article 8(1)). This kind of authorisation is valid for a provisional period of three years (a further period may be ordered) and can only be granted under certain conditions.

During the Directive's transitional period, which applies to *existing* substances not yet reviewed for inclusion in Annex I, a member state has a discretion to apply its previous national dossier requirements when reviewing authorisations for products based on these existing substances. During this period, according to article 8(2), a member state may authorise the placing on the market of a product containing an existing active substance not listed in Annex I. This period ends in 2003. Still, the Directive does require a member state to apply authorisation criteria in the review of such product authorisations (article 8(3)). In the Directive it is not clear how these criteria should be applied during this transitional period.[15]

For the regulation of the *use* of plant protection products–other than for marketing–more discretion seems to have been left to the member state. Although article 1 of the Directive states that it also concerns the use of the products, use does not fall completely within its scope. The Directive does contain some important provisions prescribing use and requiring 'proper use' of plant protection products, but it does not regulate all aspects of use, such as the training of users, requirements for application equipment and financial measures such as levies on pesticide use. Commission Regulation 451/2000 explicitly mentions the rights of member states to introduce charges, levies or fees in accordance with the Treaty (article 14). A member state does have the discretion to introduce national measures on these aspects. This discretion seems limited, though, because prescriptions on use are not to interfere with the harmonisation of pesticide marketing.

Conclusion

The centre of decision making for the authorisation of pesticides has been shifting from the member state to the EC level. On the whole, little discretion has been left with the member states to take national authorisation measures for plant protection products. Decision making on existing substances, however, has been proceeding at a slow pace. Therefore it is taking a long time for all the elements of the Directive to be realised. The authorisation criteria as well as the accompanying Uniform Principles do require that there be comprehensive protection against the adverse effects of use and do put the protection of human health and the environment at a higher priority than the objective of improving plant production, but it remains to be seen how and how fast this Directive can be put into practice.

Some changes to the Dutch Pesticides Act as a result of Directive 91/414

From an environmental and legal point of view the implementation of Directive 91/414 in the Dutch Pesticides Act can be seen as a step forward. Although the implementation period ended in July 1993, implementation did not take place until December

1994 (*Staatsblad* 1995, 4). The implementation broadened the scope of the environmental criteria laid down in article 3 of the Act and it strengthened those criteria. The Pesticides Act used to require 'reasonable certainty' that soil, water, air etc. would not be harmed in an unacceptable way. Since the implementation of the Directive it has to be 'established' that there will not be an unacceptable influence on the environment. This implies that it has become clearer that protection is to be guaranteed.

Renewals

Procedures have been clarified, especially on renewal of authorisations of products already on the market. Several Court decisions were necessary to make the consequences of the Directive clear (Vogelezang-Stoute 2000). For example, in 1998 the CBB (*College van Beroep voor het Bedrijfsleven*, the Dutch Board of Appeal for Trade and Industry, the court which hears appeals against decisions based on the Pesticides Act) ruled in the Chlorothalonil case that an application for renewal of an authorisation has to be evaluated in the same way as an application for a new authorisation. A product already on the market has to meet all the authorisation criteria to obtain a renewal of the authorisation.[16]

Renewal decisions before this decision were, in practice, often based on the application of lower standards than those applied to new authorisations. A renewal could be granted even if the information in the dossier was incomplete. The renewal would be 'administrative' or 'conditional' in that the submission of missing data would be a condition for the *next* renewal. This practice meant that an authorisation would be renewed without verifying that all the requirements for authorisation had been satisfied. Although these conditional renewals may no longer be granted, renewals can still be granted without all data being available, because the Directive provides for the possibility of 'provisional renewal'. These renewals are administrative decisions, given in situations where evaluations have not yet been completed. The Directive allows these renewals to be granted 'for the period necessary to the competent authorities of the member states, for such verification, where an application for renewal has been made' (article 4(4)). The Dutch competent authority very frequently makes use of the option to review on a provisional basis. The competent authority itself has concluded that the provisional renewal is applied in the bulk of the cases because of arrears in evaluation and decision making (CTB 2000). The President of the Board of Appeal held in several decisions that the intention was for provisional renewal to be used more strictly than the Dutch competent authority had been doing.[17] At this point one may conclude that the amendments to the Pesticides Act did not always result – at least not in the short term – in a change in the authorisation practice.[18]

Transitional measures

Since 1991 (the year in which the first Multi-Year Crop Protection Programme started in The Netherlands) the application of stricter environmental standards was postponed for quite a few widely used substances. The transitional measures were first laid down in a policy agreement for the application of the Crop Protection Programme and subsequently in a Decree.[19] These provisions, briefly stated, said that, for certain categories of products, earlier, less stringent environmental standards remained applicable for several years. However, two of the transitional provisions were declared to be non-binding, in decisions of the Board of Appeal and the President of the Board. In both decisions it was held that, as far as the provisions aimed at the non-application of the new standards, this was not in line with the Pesticides Act after the implementation of the Directive.[20]

In 2000 the prospect of non-renewal of the authorisations of products that did not meet the more stringent environmental standards had resulted in another transitional provision, one that amended the Decree and introduced new ministerial regulations. In these 'essential need' had been applied as a criterion to keep certain products on the market under certain conditions for a further period. These regulations too were declared void by the President of the Board of Appeal. These regulations could not give the jurisdiction to authorise a product without applying environmental criteria because there was no legal basis in the Pesticides Act for such jurisdiction to be given.[21]

Another effort to keep certain products that did not meet the more stringent criteria on the market was an amendment in 2001 to the Dutch Pesticides Act.[22] An 'essential need' criterion was introduced in the Act. Again, briefly stated, this implies that products based on existing substances that do not meet certain environmental and groundwater standards can be authorised in certain conditions and when there is an agricultural necessity for these products. Less stringent environmental criteria will be used here (exceeding current standards by a factor of ten to a hundred).[23] Of course, one may wonder whether this amendment is in line with the Directive, which puts agricultural interests at a lower priority than the protection of the environment. In civil proceedings initiated by environmental organisations and water companies against the Dutch government, a motion for interim relief was denied by the President of the District Court of the Hague. The President decided, *inter alia*, that it was not sufficiently clear that environmental interests were put behind agricultural interests.[24]

In the 2001 amendment of the Act a transitional provision granted several authorisations 'by law' (i.e. without the need to apply for authorisation), for a certain period, for pesticide uses considered 'essential'. These authorisations by law only lasted until the summer of 2001 because the allocated substances conditions were not met by the applicants (because of incomplete dossiers).[25]

Conclusion

One may conclude from the changes to the Dutch Pesticides Act that, despite the transitional provisions which prevented the implementation of strict environmental standards for certain substances, the implementation of Directive 91/414 on the whole has improved the quality of the Act. The broad use of the procedural renewal, however, also makes it clear that the objective to have products tested in the light of current and scientific knowledge very often in practice does not work. Either because of incomplete dossiers, or because of backlogs in evaluation and decision making, it often takes a very long time before verifications have been done for the renewal of an authorisation. During this time the product usually stays on the market. This means the central goal of the authorisation instrument, to have products properly tested according to current standards, does not seem to be realised for products already on the market.

The authorisation instrument's contribution to a more sustainable use of pesticides

There are several options in Directive 91/414 that will contribute to a more sustainable use of plant protection products.

In the first place mention should be made of the environmental principles that are implied in the provisions of the Directive and in the Uniform Principles. In these provisions the Directive provides a substantial basis for protection against adverse effects of pesticide use. In the second place, from an environmental point of view, the provision on integrated control could stimulate not only a more sustainable use of plant protection products but also a use of more sustainable products or methods. This section discusses both topics.

Environmental principles in Directive 91/414

Several of the environmental principles of the EC Treaty can be seen to be set out in the Directive and in the Annex of the Uniform Principles, such as the principle of a high standard of protection (article 2), the integration principle (article 6) and the prevention and precautionary principles (article 174(2)).

First, the *principle of a high standard of protection* is laid down in the ninth recital of the preamble to Directive 91/414. The preamble furthermore states that the protection of human and animal health and the environment should take priority over the objective of improving plant production. The EC Court of Justice referred to this recital in a case concerning the protection of groundwater. The case resulted in the stricter protection of groundwater in the Uniform Principles.[26] The high environmental protection also illustrates the integration of environmental objectives in another (e.g. agricultural) policy field (the *integration principle*).

Second, there is the *prevention principle*. A clear example of pollution prevention is the standard for groundwater. The expected concentration in groundwater has to meet the standard for drinking water or a stricter standard, when established by the Commission (Uniform Principle C.2.5.1.2). Another clear example is the criterion that the pesticide dose to be applied must be as low as possible, even where higher amounts would not result in unacceptable risks (Uniform Principle C.1.3).

Third, the *precautionary principle* is to be found in the Directive. This principle implies taking into account potentially dangerous effects if there are reasonable grounds for concern, even though scientific evidence is insufficient.[27] An example is the requirement in Uniform Principle B.1.4 to take uncertainties into account. Another element of precaution is the requirement to reject applications for which the lack of information is such that it is not possible to make a reliable decision for at least one of the proposed uses (Uniform Principle A.4).

The Directive requires that all authorisation criteria be satisfied. In line with the preamble, which gives priority to the protection of health and the environment over certain agricultural interests, the Uniform Principles do not allow that this protection be outweighed by the advantages of use. According to Uniform Principle C.1.8, a weighing of certain advantages may only take place where certain requirements concerning efficacy, impacts on plants or plant products, on vertebrates to be controlled, and on physical and chemical properties, have not been fully satisfied; the effects on human and animal health and the influence on the environment are not included in this weighing. Only in imposing conditions of use must the measures be appropriate to the expected advantages and the risks that are likely to arise (Uniform Principle C.1.1).

One may conclude that, in the long run, a good basis for protection against adverse effects of the use of plant protection products, in combination with a reasonably structured authorisation procedure, could result in products on the market meeting higher standards. This can be considered a condition for more sustainable use. However, before the use of a plant protection product may be called sustainable, more is needed.

Proper use and integrated control

As described in the introduction, essential elements of 'sustainable use' are limited application and giving priority to the least harmful method. These elements can be found in the Directive's provision on proper use and integrated control. According to the Directive, 'proper use' must be prescribed and 'integrated control' should be prescribed whenever possible (article 3(3)):

> Member states shall prescribe that plant protection products must be used properly. Proper use shall include compliance with the conditions established in accordance with Article 4 and specified on the labelling, and the application of the principles of good plant protection practice as well as, whenever possible, the principles of integrated control.

It should be noted that the term 'good plant protection practice' is not defined in the Directive. 'Integrated control' is defined (article 2(13)) as:

> the rational application of a combination of biological, biotechnological, chemical, cultural or plant-breeding measures whereby the use of chemical plant protection products is limited to the strict minimum necessary to maintain the pest population at levels below those causing economically unacceptable damage or loss.

It is clear that 'integrated control' does not only mean choosing the least harmful pesticide. It also implies using other working methods which aim to limit the use of chemical products as much as possible. When applied properly this would also mean that the necessity of use should be taken into account.

The requirement on 'proper use' is written in the imperative. Member states must prescribe proper use. One of the aims of this provision seems to be to stimulate more sustainable methods of pesticide use. Proper use, including integrated control, seems at least to be a good incentive towards a more sustainable use. Choosing the least harmful products and using these in a sustainable way (for example, at the lowest possible dose) seem inherent in this method of control.

However, Uniform Principle C.1.4 gives a very restricted interpretation of proper use:

> Member states shall ensure that decisions respect the principles of integrated control if the product is intended to be used in conditions where these principles are relied on.

According to this Principle integrated control has to be respected when a product is already 'intended to be used' in certain conditions. The scope of this Principle seems much more limited than that of the proper use article in the Directive, which requires prescribing 'whenever possible'. The words 'whenever possible' seem to be restricted in the Uniform Principles to situations where integrated control is already practised. This implies that the purpose of stimulating integrated control, which seems rather clear in article 3 of the Directive, has more or less disappeared in this Uniform Principle.

In the Dutch Pesticides Act the provision on proper use has not been implemented. For integrated control a vague provision has been established. Prescriptions have to be given 'where possible concerning the application of the principles of integrated control' (article 5(2)). In this vague provision the aim of stimulating integrated control, which can be seen in Directive 91/414, seems distant.

One conclusion can be that the Directive, by giving priority to integrated control, does have an important tool for contributing to a more sustainable use. However, this tool does not seem to be used in either the Uniform Principles or in the Dutch Pesticides Act. A more detailed definition of the elements of 'proper use' seems needed here.

Bottlenecks in the authorisation process

In general the authorisation criteria as laid down in the Directive, as discussed earlier, provide a good basis for protecting against the adverse effects of pesticide use. Environmental effects seem well covered by these criteria and by the Uniform Principles. (Although limited attention has been paid until now to airborne effects of pesticides. This is worth commenting upon, given the importance of this route. According to Uniform Principle B 2.5.1.4 the concentration in the air does have to be evaluated, but for decision making only an exposure level for operators, bystanders or workers is given, which may not be exceeded (Uniform Principle C 2.5.1.4).) Despite this good basis, there are several bottlenecks in the authorisation process that do not contribute to (and perhaps even obstruct) the goal of ensuring protection. This section describes some of these bottlenecks.

Reviewing existing pesticides according to current scientific standards

A pesticide is always authorised for a limited period only, so that at the end of this period a review can take place according to current scientific standards. In The Netherlands the maximum period for authorisation used to be ten years. Authorisations were usually given for only a few years, but within the ten-year period the authorisation could be renewed rather easily. For example, the authorisation holder would not have to apply for a renewal and the review would not be as strict as for a new authorisation. As we have seen, this changed after the implementation of Directive 91/414. According to this Directive, authorisations can still be given for a fixed period of up to ten years. A renewal, however, has to be treated in the same way as a new authorisation. It has to be applied for and the competent authority has to verify that all the criteria have been met (article 4(4)). During the verification period a provisional renewal can be granted, as we saw before. The system in the Biocides Directive is more or less the same, although in the Biocides Directive the period for granting an authorisation is more in tune with the inclusion period of the active substance (article 3(6), Directive 98/8). In the US the registration period under FIFRA used to be a maximum of five years, but in 1996 this was extended to a maximum of fifteen years.

Although one would think that with this system of periodical renewals new scientific standards would more or less automatically be applied to the existing pesticides, when time for renewal has come up, in practice special review programmes appear to be necessary to apply new standards to the products that are already on the market. And in practice these review programmes are not always successful, as the following experiences show.

As described above, the EC review of 'existing substances' is a process which is progressing at a slow pace. Until now this review process has resulted in very few definitive decisions on whether or not to list substances. It is not only at the EC level that the pace of decision making on existing substances according to new scientific standards is progressing so slowly. In the US the re-registrations of older pesticides have been going on for a very long time. The first re-registration programme dates from 1972. Because this re-registration did not work in practice, a second start was attempted in 1984. In 1988 a FIFRA amendment planned the re-registration of around 600 active substances that were on the market. Once again the planning could not be put into effect. Among the reasons mentioned for these delays were lack of finances and inefficient management of information (Ferguson & Gray (1989) for the arguments given by both the EPA and the Government Accounting Office; also Miller 1997). The 1996 amendments for the new standards for residues in food have an implementation period of ten years in total.

As already described, in The Netherlands there are also delays in the application of new standards to products already on the market. This is sometimes the case because regulations containing these new standards at the same time provide for lengthy transitional measures for ongoing applications and for pesticides already on the market, which allows for the application of older, less strict standards. Therefore the new environmental standards in a 1995 decree did not apply until 2000 to significant substances that were already on the market and did not meet the new standards. Several of these transitional provisions were declared void by the courts, often as a result of legal proceedings commenced by environmental organisations. Recently the Pesticides Act was amended to keep some of these substances on the market for a further period. In addition to these delays caused by the legislation, the application of new standards is often delayed by the authorisation practice, where provisional renewals are so often granted when the authorisation period ends and evaluations have not yet been completed.

These experiences–the review of existing substances at the EC level, the re-registration in the US and the application of new standards in The Netherlands–show how problematic the review and re-authorisation or 're-registration' can be. Sometimes this seems to result from the legislation; more often, however, the delays seem to be caused by the authorisation practice. These delays in the application of current scientific standards to the products already on the market seem rather structural. Still, an experience with re-registration in Sweden shows that a re-registration project can be completed within several years.

The Swedish review included a hazard and risk analysis, a benefit analysis and a risk benefit assessment. Cut-off criteria were applied to reject or phase out certain pesticides. (These criteria identify pesticides that are clearly unacceptable from the point of view of health and the environment or both; when an inherent property of a pesticide exceeds one of these criteria, an application will usually be rejected.) The idea behind the cut-off criteria was to facilitate prompt and easy authorisation procedures and predictability in the outcome of the decision. After the 1990–1995 review period, of the 180 'old' active substances for agricultural products, only 100 remained. Some 20 substances were removed from the market for health reasons and around 20 for environmental reasons. Many of these substances were substituted by others posing less risk. Some 15 substances were severely restricted. Many renewed approvals did get different kinds of restrictions (Bergkvist *et al.* 1996). This Swedish review process illustrates that completing a substantial review within a restricted time is not impossible, under certain conditions.

From the foregoing examples, one may conclude that the application of current scientific standards to pesticide products which are already on the market usually only takes place after a very long time and that the delays in applying new standards on review and re-authorisation seem rather structural. The history of reviewing and re-authorising shows that reviewing to current scientific standards is not only a lengthy but also an ongoing process, because new scientific insights and standards continuously appear. For example, there are new developments in the scientific insights in the field of cumulative effects and in oestrogen or other endocrine effects that pesticide residues might have. Endocrine effects, according to the FQPA, are a factor in tolerance setting under FIFRA (Sec. 408(b)(2)(D)(viii)).

Cumulative effects

In the authorisation process pesticides are evaluated one by one. Because of this, the effects of combining pesticides are not automatically taken into account. Exposure of humans and of the environment to pesticides is often exposure to a combination or mixture of pesticide residues. Exposure will often be an aggregate exposure, through different routes or media, such as food, drinking water and air. These mixtures and aggregate exposures, involving various compounds and different media, can produce synergistic effects. The regulatory focus on single pesticides and single effects creates an essential gap in the authorisation process. A cumulative risk assessment would seem appropriate here (Wargo 1998). Wargo gives an interesting analysis of the relation between science and regulation: 'Moreover, this regulatory focus on single compounds rather than complex mixtures reveals how narrow science can drive narrow regulation' (Wargo 1998).

Unlike Directive 91/414, Directive 98/8 made a start by addressing in its text one aspect of the problem of 'cumulation effects'. A requirement for listing an active substance is that, where relevant, the cumulative effects of the use of biocidal products containing the same active substances have to be taken into account (article 10 (1)). Note that this Directive refers to *the same* active substances. Therefore the cumulation that has to be taken into account seems more restricted than under FIFRA. In fact, what is called 'cumulation' under the Biocides Directive seems more related to the aggregate exposure as applied under FIFRA, as mentioned below. It remains to be seen how cumulation effects will be taken into account in future decision making under the Biocides Directive.

In the US the 1996 FQPA introduced a cumulative risk assessment for pesticides by requiring the EPA to take into account cumulative effects when registering or re-registering pesticides under FIFRA. Here it is the cumulative effects of pesticides with a common mechanism of toxicity that have to be considered. These requirements are laid down as FIFRA residue requirements.[28] The focus of EPA during the past years, in developing a cumulative risk policy, has been on the organophosphate insecticides (Miller 1997). An evaluation of the first years is given by the Consumers Union of United States (2001). Under FIFRA, in addition to these cumulative effects, aggregate exposure to a pesticide chemical residue (the same pesticide through various exposure routes) also has to be taken into account when setting safe tolerance levels.[29]

The developments under FIFRA and the FQPA do not mean the problems of cumulation and aggregate exposure are solved. One of Wargo's conclusions is that the law governing pesticides is still fractured chemical by chemical, product by product and medium by medium. The result of this is a knowledge of the risk, one chemical at a time and one effect at a time (Wargo 1998: 304). As we saw above, it can take many years before new scientific insights are incorporated in practice into the authorisation process. This will probably also be the case for the cumulative effects and aggregate exposure, for which the first steps are now being taken.

Substitution

One of the effects of the product-by-product approach in the authorisation process is that the comparison with other products hardly plays any role and that the 'substitution principle' is not applied in the decision making process. This principle means that the authorisation of a substance or product can be ended or rejected if a less hazardous substance or product is available for the same use.

In EC legislation the substitution principle was already mentioned in Directive 79/117/EEC prohibiting the placing on the market and use of pesticides containing certain active substances. The tenth recital in the Preamble to Directive 79/117 states that certain national derogations to this Directive were to be phased out as soon as less hazardous treatments became available.

Directive 91/414 does not have the substitution principle as part of its authorisation criteria. However, in the Directive's provisions on integrated control a substitution approach seems to be implied. The directive circumscribes integrated control as 'the rational application (...) whereby the use of chemical plant protection products is limited to the strict minimum necessary to maintain the pest population at levels below those causing economically unacceptable damage or loss' (article 2(13)). Applying the substitution principle seems inherent to this way of using pesticides, while integrated control aims at an application which strictly minimises the use of plant protection products.

The Biocides Directive in 1998 introduced a substitution criterion for the evaluation and decision making on the active substances for biocides. The Directive contains a comparative assessment for the application of this criterion (article 10(5)). A discretion has been given about whether to refuse to add an active substance to the list, or to remove it from the list:

(...) if there is another active substance on Annex I for the same product type which, in the light of scientific or technical knowledge, presents significantly less risk to health or to the environment (...)

adding that:

When refusal or removal is considered, an assessment of an alternative active substance shall take place to demonstrate that it can be used with similar effect on the target organism without significant economic and practical disadvantages for the user and without an increased risk for health or for the environment.

As one can see from this last quote, a thorough assessment is required. Further conditions are also laid down, such as the requirement that the diversity of active substances has to be adequate in order to minimise the occurrence of resistance in the target organism. It remains to be seen how the substitution principle will be applied in the listing of the active substances for biocides.

In the above-mentioned Swedish re-registration project the substitution principle was applied. In Sweden the substitution principle is laid down in the Swedish Act on Chemical Products.[30] Here substitution is considered a precautionary matter:

> Anyone handling or importing a chemical product must take such steps and otherwise observe such precautions as are needed to prevent or minimise harm to human beings or to the environment. This includes avoiding chemical products for which less hazardous substitutes are available.

Denmark (which also uses the substitution principle in its authorisation process) and Sweden have proposed implementing the principle into Directive 91/414 both for the decision making on active substances and for authorising plant protection products (Bergkvist 1998). Unlike these member states, in The Netherlands the substitution principle was removed from pesticide legislation before it could be applied. It was brought into the Pesticides Act in 1993 and removed again in 1994 with the implementation of Directive 91/414.

One may conclude that, although the substitution principle is a rather new element in the product-by-product approach of the authorisation of pesticides, experiences in Sweden and Denmark show that application of this principle is not impossible. The use of this principle may offer a chance to give priority to less hazardous products. This is an opportunity that is missed under Directive 91/414. One could argue that the requirements of high protection and the requirements on integrated control imply the application of a substitution approach. However, a necessary condition for the use of the principle – a procedure for comparative assessment – is lacking.

Concluding remarks

The bottlenecks described in this paragraph (i.e. the delays in the application of current scientific standards, not taking into account cumulative effects and the absence of the substitution principle) can be seen as serious shortfalls in the authorisation process as laid down in Directive 91/414, not only from an environmental point of view but also in the light of the goals of the Directive regarding the high protection standard. A different approach to decision making, as well as an amendment of the Directive, seems required to remove these bottlenecks.

There are many other aspects that complicate the current authorisation process and that might contribute to the adverse effects of pesticides use. Examples are shortcomings in the use of information from the 'open literature' (i.e. data which are publicly available), the lack of monitoring data concerning the effects of use and, last but not least, lack of information for the pesticide user. It is beyond the scope of this chapter to deal with all these aspects.

Some of these aspects could probably be turned, within the current system, into opportunities for a more sustainable use of pesticides. An authorisation could, for example, have as a condition an obligation on an authorisation holder to inform the buyer adequately about environmental effects of use. It is quite amazing that such a requirement does not now exist. Neither the government nor the competent authority (at least, not in The Netherlands) provides adequate and up-to-date information for the non-professional consumer on environmental effects, to give just one example.

Final conclusions and EC policy developments

The authorisation system as laid down in Directive 91/414, with its stringent authorisation criteria, can be considered a good basis for reaching a high standard of protection against the adverse effects of the use of plant protection products. The implementation of the Directive in the Dutch Pesticides Act, for example, did improve the legal and environmental quality of this Act.

However, in the evaluation and decision making required by the authorisation process, there are several bottlenecks that seem to prevent the realisation of a high standard of protection. One of the main goals of authorisation (i.e. that marketed products are tested according to current scientific standards) is often not realised because of delays in reviewing authorisations of existing pesticides. These delays often seem rather structural. This is the case for several review and re-authorisation processes at the national and the EC level, and also at the federal level in the United States.

Because of the product-by-product approach, the cumulative effects of combining pesticides and the aggregate effects of exposure by different routes have so far barely been integrated into the evaluation and decision making process. The introduction of these effects into the decision making process is only just starting. This is also the case in the use of the substitution principle. By the application of this principle, giving priority to less hazardous products or methods, the authorisation instrument could contribute to a more sustainable use of pesticides. In combination with the provisions on proper use and integrated control, this contribution could probably be substantial. So far, however, it seems the proper use and integrated control provisions are not being applied in practice.

In 1998 the Commission organised a workshop on the sustainable use of plant protection products in the EU. The workshop was based on several studies of possibilities for an additional (environmental) EC Policy on plant protection products. A Communication of the Commission on a sustainable use of these products was announced in 1998 (Proceedings 1998), but was not issued until 2002.[31]

Recent developments at the EC level show that, since the effects of a high standard of protection are becoming increasingly clear, policy priorities seem to be shifting more towards agricultural protection. In Commission Regulation 451/2000 an 'essential use criterion' is introduced.[32] The Commission may, under certain conditions and on a case-by-case basis, take temporary measures on product use if an essential need has been demonstrated and an alternative is lacking. By the autumn of 2001 preparations for the decision making on this point were taking place. Member states are 'collecting' their essential uses.

The temporary measures referred to by Regulation 451/2000 can only take place as part of the overall decision making on the transitional period for Directive 91/414. This period will be extended to 2008. This will be part of the evaluation and amendment of the entire Directive, based on the progress report by the Commission to Parliament and Council (COM (2001) 444 final). It will be interesting to see what the role of the European Parliament will be in this matter. The changes to the Directive, which the Commission is suggesting in its report, will have to be decided on by the Council in co-operation with Parliament.[33] One of the main issues here should be how measures concerning 'essential use' can be brought in line with the priority given by the Directive to a high standard of protection over the objective of improving plant production.

Acknowledgements

The author thanks Professor Hanna Sevenster and Greg Korbee (BA JD) for their very useful comments on an earlier draft of this chapter.

Notes

1   Council Directive 91/414/EEC of 15 July 1991, *OJ* 1991 L 230/1.
2   Since this text was finalised (October 2001), important legislative and policy developments have taken place, which could not be integrated into the text. References to these developments have been added in the endnotes.
3   Wet bestrijdingsmiddelen en meststoffen, Staatsblad (*Stb.*) H 123, 1947.
4   Bestrijdingsmiddelenwet 1962, *Stb.* 1962, 288.
5   Parliamentary Documents TK 1970–1971, 11 262, nr. 3, p. 6.
6   Federal Insecticide, Fungicide and Rodenticide Act, 25 June 1947, Pub. L. No. 86-139, 61 Stat. 163 (1947).
7   Food Quality Protection Act, Pub. L 104–170, 110 Stat. 1489 (1996).
8   Directive 98/8/EC of the European Parliament and of the Council of 16 February 1998, *OJ* 1998 L 123/1.
9   Examples are Directive 90/642/EEC, *OJ* 1990 L 350/71 (later amended) on maximum levels for pesticide residues in and on certain products of plant origin, including fruit and vegetables; Directive 1999/45/EC, *OJ* 1999 L 200/1 on classification, packaging and labelling of dangerous preparations; Directive 79/117/EEC, *OJ* 1978 L 33/36 (later amended) prohibiting the placing on the market and the use of pesticides containing certain active substances; Directive 76/769/EEC, *OJ* 1976 L 262/201 (later amended) on the marketing and use of certain substances and preparations; Regulation EEC/2078/92, *OJ* 1992 L 215/ 85 (later amended) on agricultural production methods compatible with requirements of the protection of the environment.
10 Commission Regulation (EEC) 3600/92, *OJ* 1992 L 366/10, later amended, laying down the detailed rules for the implementation of the first stage of the programme of work referred to in article 8(2) of Council Directive 91/414/EEC. Commission Regulation (EC) no. 451/2000 of 28 February 2000, *OJ* 2000 L 55/25, laying down the detailed rules for the implementation of the second and third stages of the work programme.
11 Commission Regulation (EC) 933/94, *OJ* 1994 L 107/8.
12 Commission Regulation (EC) 2266/2000, *OJ* 2000 L 259/27, amending Regulation 3600/92.
13 Commission Regulation 451/2000, article 6.

14 Except for confidential information in the meaning of article 14 of Directive 91/414, e.g., Commission Directive 98/47/EC, *OJ* 1998 L 191/50, including azoxystrobin.

15 Decision, European Court of Justice 3 May 2001 (case C-306/98) does not bring much clarity to this matter, unlike the opinion of the Advocate General (21 September 2000) in this case.

16 CBB 29 January 1998, *Milieu en Recht (M en R)* 1998/4, nr. 33 m.nt. Vogelezang-Stoute, *Administratiefrechtelijke Beslissingen (AB)* 1998, nr. 111, m.nt. JHvdV, *Nederlands Tijdschrift voor Europees Recht* 1998, pp. 120, 121 (Sevenster).

17 CBB 29 January 1998, and CBB 15 July 1999, *M en R* 1999/12, nr. 121, m.nt. Vogelezang-Stoute (metamnatrium).

18 In 2001, in The Netherlands, the provisional authorisations became a structural part of the authorisation practice, by the introduction of a so called 'priority list'. In 2002 the Board of Appeal held that the competent authority did not have the discretionary power to decide on the basis of this priority list, which resulted in some 150 provisional authorisations (CBB 2 July 2002, AWB 01/722). In the meantime the Pesticides Act had virtually been amended again, to create a legal basis for the 'listing' (Parliamentary Documents 2001–2002, 27 085).

19 *Bestuursovereenkomst uitvoering meerjarenplan gewasbescherming* (1993); *Besluit milieutoelatingseisen bestrijdingsmiddelen (Stb.* 1995, 77, later amended).

20 CBB 29 January 1998 and President CBB 11 May 1990, *AB* 1999, 331, m.nt. JHvdV, *M en R* 1999/7/8, nr. 69, m.nt. Vogelezang-Stoute) (Dichlorovos case). The transitional measures were article 9 and article 8(1)(a) *Besluit milieutoelatingseisen bestrijdingsmiddelen.*

21 President CBB 10 July 2000, *AB* 2000, 320, m.nt. JHvdV, *M en R* 2000/3, nr. 38, m.nt. Vogelezang-Stoute (*Tijdelijke regelingen*).

22 *Stb.* 2001, 68, adding a new article 25c.

23 *Regeling toelatingseisen landbouwkundig onmisbare gewasbeschermingsmiddelen, Stb.* 2001, nr. 42.

24 Pres. Rb. den Haag 30 May 2001, *KG* 2001/565 and *M en R* 2001, 7/8, nr. 158K.

25 In the 2002 amendment of the Pesticides Act these 'authorisations by law' were laid down in the Act. Products based on allocated substances will be considered authorised and will not be reviewed until an EC review has taken place (Stb.2002.461).

26 Case C-303/94, Parliament v. Council, ECR 1996 I-2943.

27 Communication from the Commission on the precautionary principle, COM(2000)1.

28 FDCA Sec. 408(b)(2)(D)(v).

29 FDCA Sec. 408(b)(2)(A)(ii).

30 SFS 1985:426, section 5.

31 The Communication from the Commission was published in 2002: 'Towards a Thematic Strategy on the Sustainable Use of Pesticides', COM(2002) 349 final. In the framework of the 6th Environmental Action Programme the strategy sets out objectives to minimise hazards and risks, improve controls, reduce levels of harmful active substances, encourage the use of low-input or pesticide-free crop farming and establish a transparent system for reporting and monitoring. NGOs have proposed a more stringent approach, with a text for a Directive on pesticide use reduction, complementing Directive 91/414 (Pesticides Action Network Europe and European Environmental Bureau, May 2002).

32 *OJ* 2000 L 55/25 (article 15).

33 Environmental Council conclusions were adopted on 12 December 2001. The European Parliament Resolution on the Commission Report was adopted on 30 May 2002. The resolution contains strict conditions for the extension of evaluation deadlines and supports the introduction of the substitution principle, among many other substantial calls for change.

References

Bergkvist, P. (1998) Swedish and Danish paper on the use of the substitution principle in regulation of plant protection products. National Chemicals Inspectorate, Solna.

Bergkvist, P., Bernson, V., Jarl, S. & Törenlund, M. (1996) Reregistration of pesticides in Sweden: Results from the review 1990–1995. *Pesticide Outlook*, 7 (1), 12–18.

Cardonnel, P. & van Maldegem, K. (1998a) The Biocidal Products Directive. *European Environmental Law Review*, (11), 261–8.

Cardonnel, P. & van Maldegem, K. (1998b) The Biocidal Products Directive. *European Environmental Law Review*, (12), 315–21.

Consumers Union of United States (2001) *A Report Card for the EPA. Successes and Failures Implementing the Food Quality Protection Act.* CUUS, New York.

CTB [*College voor de toelating van bestrijdingsmiddelen*] (2000) *Notitie definitieve prioriteitstelling*, December 2000, p. 2. [http://www.ctb-wageningen.nl/priobew.html, 24–09–01]

Ferguson, S. & Gray, E. (1989) FIFRA Amendments: A Major Step in Pesticide Regulation *Environmental Law Reporter*, 19, 10070–10082.

Formica, M.C. & Miller, M.L. (1999) Legislative Background of FIFRA. In: *Pesticides Law Handbook* (ed. M.L. Miller), pp.283–93. Government Institutes, Rockville, MD.

Miller, M.L. (1997) Pesticides. In: *Environmental Law Handbook*, (ed. T.F.P. Sullivan), pp.284–327. Government Institutes, Rockville, MD.

Proceedings (1998) Proceedings of the Second Workshop on a Framework for the Sustainable Use of Plant Protection Products in the European Union, Brussels, 12–14 May 1998. [http://europa.eu.int/comm/environment/ppps/workshop.pdf, 17/10/2001]

Vogelezang-Stoute, E.M. (1999) European Community legislation on the marketing and use of pesticides. *Review of European Community & International Environmental Law*, (2), 144–151.

Vogelezang-Stoute, E.M. (2000) Directive 91/414/EEC and its implementation in the Dutch Pesticides Act *European Environmental Law Review*, 9. 8/9, 237–42.

Wargo, J. (1998) *Our Children's Toxic Legacy. How Science and Law Fail to Protect Us from Pesticides.* 2nd edn. Yale University Press, New Haven and London.

Chapter 4
Innovation in the Agrochemical Industry

Frank den Hond

Introduction

Companies in the agrochemical industry are science-based, multibillion-dollar, multi-national companies that develop and produce pesticide products in an increasingly global marketplace. They make their money by providing 'crop protection solutions' to farmers. Thus, they also develop and market related agricultural inputs, such as seeds and transgenic crops (as with Monsanto, Syngenta, Aventis) and fertilisers (e.g. BASF). But their societal mission is much broader; it encompasses contributing to sustainable development by providing tools and services that allow farmers to produce food and fibres for a growing world population, whilst minimising the associated risks to human health and the natural environment. For example, Aventis CropScience directs its:

> research and development efforts ... in crop production ... towards the creation of products that support efficient and sustainable agricultural production practices. These innovations are enabling farmers to produce high-quality, cost-effective yields for the benefit of stakeholders, consumers and the environment. At this very moment, advances in biotechnology, including genetically improved seeds, are helping us to deal with the increasingly complex challenges in regards to food safety, quality and supply.[1]

Monsanto – recently taken over by Pharmacia – has formulated its business purpose and mission as:

> abundant food and a healthy environment ... . For the world's food producers, we work to deliver products and solutions to help them reach their goals in ways that: meet the world's growing food and fiber needs; conserve natural resources; improve the environment.[2]

Further, Syngenta aims 'to be the leading provider of innovative solutions and brands to growers, and to the food and feed chain'.[3]

The persistence of the pesticide problem stands in sharp contrast to the agrochemical industry's claim of being highly innovative and socially responsible. Because making social responsibility operational and measuring it is notoriously difficult, and because the industry's claim of social responsibility is based on the marketing of innovative products, the objective of this chapter is to discuss the industry's innovativeness rather than its social responsibility. Thus, our questions are: How can innovation processes in the agrochemical industry be characterised? What are the drivers and outcomes? The relevance of such questions is to enable speculation about the question of whether (and how) the manner in which the agrochemical industry is innovative with regard to active ingredients might be related to the persistence of the pesticide problem. Because of this focus on innovation, other players in the industry, including the producers of generic, off-patent active ingredients, companies that specialise in formulating and trading pesticide products, and companies developing the application hardware, are only considered if relevant to the argument.

The second and third sections point out the process of innovation in the agrochemical industry by discussing the recent discovery and product development of a specific group of pesticides, the strobilurin fungicides. By focusing on working hypotheses, preliminary conclusions and contingencies, we can identify 'research lines' and 'critical research events' (Vergragt 1988) which, in turn, allow us to develop operational-level insight in the research processes that underlie the industry's attempts to solve the crop protection problem. The innovation process is interpreted from an evolutionary perspective by making use of the notions of technological trajectories and technological paradigms (Dosi 1982; Nelson & Winter 1982).

The chapter continues by characterising the drivers behind, and strategic aspects of, innovation in the agrochemical industry. As in the pharmaceutical industry, competition in this industry is predominantly about the introduction of new products (Hartnell 1996), rather than the cost levels of production processes *per se*, as could be expected from the increasing importance of producers of generic, off-patent pesticide products. It appears that the industry has reached a mature state of development in which ever-increasing scales of operation are crucial for competitiveness, but also that renewal of crop protection concepts might be warranted from market and social perspectives. We make use of radicality of innovations and related concepts in order to interpret in the fifth section *how* the industry is innovative. Innovations can be incremental or radical at either the component or systems level (Henderson & Clark 1990), because of their competence enhancing *versus* competence destroying character (Tushman & Anderson 1986). They may turn out to be industry sustaining or industry disrupting technologies, depending on whether or not they relate to customers in traditional, established ways or in entirely news ways, perhaps even attracting a new customer base (Christensen 1997).

Finally, the sixth section reviews the outcomes of innovation processes in this industry by analysing the availability of pesticide products in The Netherlands over a period of almost 35 years from 1965 until 1999. In this way, questions of how quickly and to what extent innovative pesticide products penetrate the crop protection market may be addressed.

Strobilurin fungicides: research lines and critical research events[4]

Synthetic fungicides have been widely used for the protection of agricultural crops since World War II. Early synthetic fungicides were contact fungicides; their effectiveness depends on the equal and complete distribution of the compound on the leaves and stems of crop plants. The agricultural use of synthetic fungicides was greatly enhanced when in the late 1960s systemic fungicides were commercialised. As opposed to contact fungicides, systemic fungicides are absorbed by the crop and distributed by active transport to all parts of the plant, thus providing a much more equal protection of the plant, including non-treated parts and new shoots and leaves. The discovery of various groups of systemic fungicides marked the beginning of a new period of anti-fungal crop protection. However, the extensive agricultural use of systemic fungicides led to an increasing number of reports on reduced functionality of the active ingredients because fungal diseases developed resistance to commonly used fungicides. The first instances of resistance were recorded in the early 1970s (Heaney *et al.* 1994). Resistance spread so widely that already in the mid-1990s it was concluded that 'the performance of most of the modern, systemic fungicides has been affected to some degree' (Brent 1995).

For various reasons, including the resistance problem, a need was felt by the late 1970s for fungicide products based on new active ingredients with new modes of action. Agrochemical companies started to look for new classes of fungicidal compounds in order to be able to differentiate themselves from the product offerings of their competitors. Several (groups) of fungicides were discovered and have been commercialised since the late 1980s and early 1990s.

In the early 1980s, the opportunity was recognised to make use of naturally occurring compounds (toxins) from plants, fungi, bacteria, etc. for crop protection purposes. Either the compounds themselves could be used when available or producible in sufficiently large quantities, or they constituted useful starting points ('leads') for the synthesis of analogues with improved biological and physical properties (Beautement *et al.* 1991), as well as greater opportunity for patent protection. Various agrochemical companies started programmes to evaluate natural products as potential leads for new synthetic pesticides in the early 1980s. Natural products were selected on the basis of literature studies, the isolation of natural compounds and the subsequent elucidation of their structure, and the consultation of external sources such as academic research groups. An interesting example of fungicide products that are analogues from natural lead compounds are the strobilurin fungicides. BASF and ICI worked simultaneously, yet independently, on this new class of fungicides. They have marketed strobilurin fungicides since the mid-1990s. Other companies followed suit.

Strobilurin fungicides are the synthetic analogues of naturally occurring β-methoxyacrylates, including strobilurin A, which is the most simple structure in this class of compounds (compound **1** in Fig. 4.1). Such compounds are isolated from several genera of basidiomycete fungi. Presumably, the biosynthesis of these fungicidal compounds gives them an advantage over other fungi in the competition for nutrients. Strobilurin A was first isolated by a group of Czech academic researchers in 1969; they observed the compound's fungicidal activity and published its (correct) chemical structure in a patent in 1979. Independently from the Czechs, a group of German academic scientists started to work on biologically active metabolites from basidiomycete fungi in 1975. Among the first metabolites they isolated was strobilurin A; the Germans, too, observed a remarkable fungicidal activity but they were mistaken in their characterisation of its chemical structure. The two compounds were not recognised as the same, because of a wrong description of physical properties by the Czech researchers. The resulting confusion was resolved in the mid-1980s. Following their discovery, the German group synthesised the natural product as well as some 30 analogues and derivatives and they tested them for anti-fungal activity. By comparing the structure and the antifungal activity of each of these compounds, they were able to conclude that the terminal β-methoxyacrylate group is the compound's toxophore.[5]

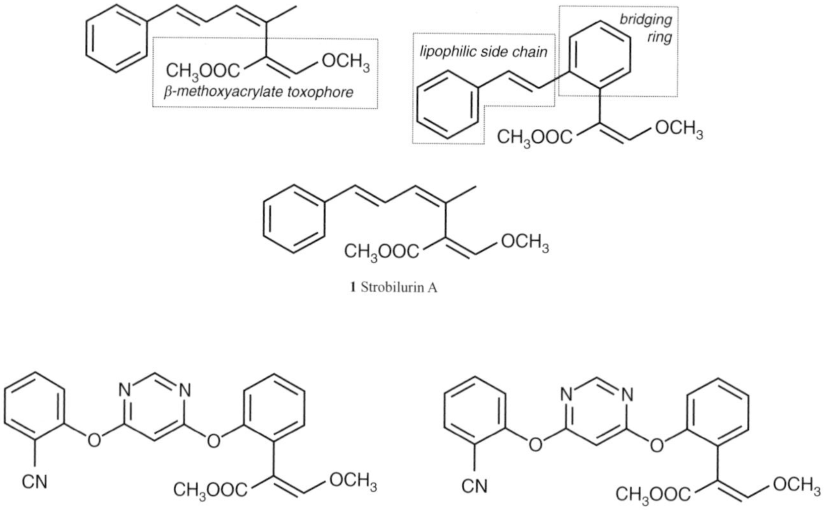

2 ICI A5504 "Azoxystrobin"

3 BAS 490F "Kresoxim-methyl"

Fig. 4.1 Chemical structures of strobilurin A and related fungicides.

In the early 1980s, the papers by the German research group provoked interest in the chemistry of strobilurin derivatives at the ICI laboratories. ICI learned from them that the various β-methoxyacrylates are structurally related and that they share the same mode of action. Moreover, none of the fungicides sold at that time exhibited a similar mode of action, so cross-resistance between β-methoxyacrylates and other classes of fungicides was unlikely to occur. This was considered most relevant given the increasing problems with resistance. Finally, because the mode of action was known, an *in vitro* assay for the compounds that were targets for synthesis could be established. 'This, therefore, seemed to be a very attractive starting point for synthesis' (Beautement *et al.* 1991). In developing crop protection chemicals from the strobilurin chemistry, ICI followed a step-by-step approach. Three steps can be distinguished: (1) confirming the reported fungicidal activity of strobilurin A and other β-methoxyacrylates, (2) a chemical synthesis programme which led to the discovery of more potent, stable, and systemically active compounds, and (3) the optimisation of the systemic activity of these compounds which culminated in selecting ICI A5504 (compound **2** in Fig. 4.1) as the active ingredient for Zeneca's new strobilurin fungicides. During these steps ICI/Zeneca tried also to find new, more effective toxophores. Optimisation of formulations and preparing for production followed the selection of ICI A5504 for commercialisation. ICI first presented ICI A5504 at the 1992 Brighton Crop Protection Conference. The objective was to make it the leading broad-spectrum fungicide. The first registration was obtained in Germany in April 1996; registrations in other countries quickly followed. However, the development process did not end at this stage. Patent applications describing related compounds were filed, for various reasons. There were still risks of failure, for example the risk that ICI A5504 could not be mixed into a stable and convenient formulation, or the risk that the synthesis of ICI A5504 could not be brought to full-scale production within the cost limits set by the company. Therefore, alternatives had to be available. Zeneca decided to stop further development work by the end of 1996, shortly after the start of full-scale production of ICI A5504. There was no longer a need for alternatives to ICI A5504. Synthesis work in the strobilurin area continued at a smaller scale and was mainly aimed at the re-synthesis of the compounds reported by competitors in the patent literature.

Like other agrochemical companies, BASF, too, started a programme to evaluate natural products as potential leads for new synthetic pesticides in the early 1980s. They increased their in-house research capacity for the isolation and characterisation of natural compounds by various research co-operations with universities in Germany and abroad, including the German group of researchers mentioned before. In July 1983, BASF received a sample of strobilurin A from this group. In testing its reported fungicidal activity in a greenhouse setting, BASF only found disease control at relatively high dose rates. Knowing the compound's chemical structure, it was hypothesised that strobilurin A is sensitive to quick photolytic, chemical or metabolic breakdown. Thus, BASF inferred, the structure of the compound needed to be stabilised. Consequently, BASF synthesised many different compounds by varying the backbone and side chain of strobilurin A. BASF considered very interesting for further testing a number of these compounds, but much to their annoyance they found that this was effectively blocked by an ICI patent application. This situation provoked considerable discussion with the fungicides researchers on whether or not to continue the project. The very promising results from greenhouse screening and small field plots–very broad scope of activity including against some very hard-to-control fungal diseases and high levels of activity–were convincing arguments to continue research investment, but the crucial question was whether chemical variations would give sufficiently active compounds outside of the ICI claims. As BASF had already started to work on modifying the toxophore, they decided to continue research work in this direction. BASF systematically modified the toxophore and tested for activity, but the reported results indicated that, in addition to the original toxophore of strobilurin A, only an oxime ether toxophore exhibited high levels of fungicidal activity. One of the many compounds synthesised and tested was BAS 490F (compound **3** in Fig. 4.1); the choice of BAS 490F as the most promising development product was guided by the results of field tests, the favourable eco-toxicological profile and considerations about the potential cost of production. BAS 490F is 'quasi-systemic' as it diffuses by passive transportation over the foliage surface where it inhibits the sporulation of pathogenic fungi. The researchers involved stress that 'this particular combination of characteristics [of BAS 490F] has not been designed *a priori* as an objective and then developed in a straightforward manner' (Sauter *et al.* 1995). BAS 490F was also presented at the 1992 Brighton Crop Protection Conference.[6] BASF, obtained its first registration for BAS 490F in March 1996. The product is registered now for various applications in almost all European countries, and many overseas countries. Large-scale production started in September 1995.

Other companies, observing patent activity from both ICI/Zeneca and BASF and considering that this field might be a new and promising area for developing fungicides, joined the strobilurin bandwagon. In most instances, they first reproduced the reported results, and secondly started to search for 'gaps' in the ICI and BASF patent applications in order to exploit research directions that were not covered by these patents. However, this research often did not lead to significant improvements in activity when compared to the activity as observed from the competitors' compounds. Patent applications reveal that an increasing number of competitors had become active in the field, which led follower companies to consider that the chances of finding new side-chain variations in time for patent protection would be very low. Hence, many of these companies decided to expand their research strategies into modifying the toxophore, at times using molecular modelling in order to find toxophores that resemble the methoxyacrylate group in size, polarity or electronegativity or both. If a company was lucky enough to find a new biologically active toxophore that was not yet described in one of its competitors' patent applications, it might be able eventually to develop a marketable pesticide product.

The similarities in the strobilurin research at ICI and BASF are striking. Both companies started to work on this chemistry in the early 1980s. They adopted very similar research strategies, formulated similar hypotheses and obtained the same results. As Sauter *et al.* (1995) say: 'These parallel findings are certainly among the most striking examples of how independently generated ideas finally result in similar, if not identical, developments in an interesting and highly competitive field such as the strobilurins.' ICI and BASF scientists had no direct contacts with each other on the strobilurin chemistry, nor were they aware of each other's development activities until this was revealed by patent applications.

Trajectories, heuristics and a technological paradigm

The account of the discovery, research and development activities within the area of strobilurin fungicides illustrates the emergence of technological trajectories in the agrochemical industry in general. A 'technological trajectory' can be understood as the 'direction of advance' within a particular technological paradigm (Dosi 1982). It emerges over time by the application of specific heuristics, which is a principle that is believed to contribute to reducing the search for particular solutions (Nelson &

Winter 1982). Technological trajectories and paradigms have a mental component of what are perceived to be acceptable and adequate directions for search (see Spender 1980), and a physical dimension, of what nature tells the scientist are plausible hypotheses and feasible solutions. In the trajectory of strobilurin chemistry, the structuring over time of the innovation process closely resembled Rogers' (1995) classical description of a linear process of generation and diffusion of innovations. However, in contrast to Rogers' model, there was not so much of a specific initial problem identification stage.[7] Comparison with other studies of agrochemical innovation, e.g. Achilladelis *et al.* (1987), den Hond (1998b) and Hartnell (1996), suggests that this is a common, if not the dominant, innovation pattern in the agrochemical industry. Innovation in this industry is thus largely 'technology push', rather than 'market pull'.

The strobilurin account also contains many elements of what could be considered the technological paradigm of the agrochemical industry. A 'technological paradigm' can be understood as the industry-wide and shared understanding of what is a 'normal' pattern of solving particular problems; it embodies strong prescriptions on which directions of technical change to pursue and which to neglect. A technological paradigm thus has a strong exclusion effect in making R&D managers, scientists and engineers 'blind' to other solutions to the particular problem at hand (Dosi 1982). In this respect, it is close to what Spender calls 'industry recipes', i.e. 'the shared knowledge-base that those socialised into an industry take as familiar professional common sense' (Spender 1980). Industry recipes emerge as managers develop ways of dealing with uncertainty, often by imitating the earlier, creative solutions or heuristics of other companies in their business, in this case relating to agricultural and societal demands for crop protection.

Because the 'technology push' model of innovation has been criticised for its lack of consideration of market signals and broader societal concerns, the question arises of how market demand for crop protection is being addressed by the agrochemical industry. The discovery and the subsequent development and marketing of new synthetic chemicals are technical 'solutions'. But what are the problems? And how are those problems related to the solutions that are being developed?

Among the problems are (Hartnell 1996):

- competitive pressures within the industry that require agrochemical companies continuously to market new pesticide products, preferably with new modes of action;
- a market demand for methods for agricultural crop protection (although not necessarily restricted to chemical control);
- the development of resistance to specific (groups of) pesticide products by pest populations (a 'second-order' problem because of the continuous use of a limited number of pesticides all having the same mode of action).

Within the agrochemical industry, the dominant solution to all these problems is to identify new chemical compounds that exhibit useful biological activity and that may be starters ('leads') for a synthesis programme to optimise their physical, chemical and biological characteristics. However, the various elements that constitute the technological paradigm of agrochemical innovations are not stable. Over time, incremental changes have occurred in the applied heuristics and in the selection environment. Moreover, the selection environment may pose conflicting demands. It will be argued in the following that market demand and social concerns only constitute weak and indirect influences on the technological paradigm and its associated trajectories and heuristics.

The isolation of natural compounds is one of various sources that companies use to identify new compounds as potential leads for further chemical synthesis. 'To look for naturally occurring compounds as potential leads' is a heuristics. Other heuristics include 'at random' screening of chemicals originating from a variety of sources, 'analogues chemistry' in which the aim is to invent around the compounds patented by competitors, and 'biorational design' (Evans & Lawson 1992). In biorational design the idea is to design dedicated molecules for inhibiting specific metabolic processes in the target organism (Schwinn & Geissbühler 1986). At random screening of the by-products from the chemical industry and the analogues approach were the dominant heuristics until well into the 1970s (Achilladelis *et al.* 1987). After that period, additional heuristics, such as natural compounds and biorational design, were promoted, albeit with mixed success. Whereas several cases of the successful introduction of new active ingredients based on natural leads are known, no successful case of pesticide innovation using the biorational design approach has been reported (Hartnell 1996), unless the definition of 'biorational' is expanded to include the changing of metabolic processes in crops by the new biotechnology in order to make them tolerate or produce specified active compounds. Recent developments in chemical synthesis and screening for biological activity have again put 'at random' screening at the forefront of pesticide innovation. Automated synthesis of very large numbers of compounds through combinatorial chemistry and the development of *in vitro* assays for screening compounds at high throughput rates – both originally developed in the pharmaceutical industry – introduced economies of scale in the identification of potential leads. By the late 1990s, the larger agrochemical companies were able to screen about 50 000 compounds per year and the aim was to double this within subsequent years. Thus, at random screening is regaining importance over the more focused and 'intelligent' search modes of natural products and biorational design. When compared to the heuristics of at random screening during the earlier decades, the 'turn to Nature' represented a shift in heuristics, which has returned to at random screening but is now made more potent by innovations in chemical synthesis and screening.

Regarding the selection environment, the need to have pesticide products registered prior to sales is a critical element. The criteria for the registration of pesticide products have been modified and renewed constantly since their introduction. Initially, criteria were introduced which prohibited adverse effects on consumers, agricultural workers and livestock, whereas later, increasingly stringent criteria were introduced to prevent damage to non-target organisms and the natural environment at large. Although it has been suggested that those companies who are able to meet these criteria in new product development may gain a relative competitive advantage over those companies with relatively older product portfolios (Paulino 1997), the regulatory process has not had a direct influence on the choice of which heuristics to adopt. The criteria function as 'cut-off' values in deciding whether or not to continue developing specific compounds. They are part of the selection environment for pesticide innovations and, as such, agrochemical companies anticipate the selection process by submitting for regulatory review only those new active ingredients that are likely to pass all the criteria. Rather than looking for chemical structures that are 'optimised' for pest control, they are satisfied by meeting the regulatory demands.[8]

The selection environment may pose conflicting demands, especially in the agrochemical industry. For example, the pesticide innovator finds himself trapped between the need to reduce environmental impact and the need to secure effective biological activity (Stetter 1993). Thus, the decisions to commercialise ICI A5504 and BAS 490F respectively reflect the compromise reached between the various conflicting demands. 'Optimisation' of the chemical structure implies developing satisfactory performance levels at various dimensions. The fact that BAS 490F is a quasi-systemic, rather than a systemic, compound is illustrative evidence of this satisfying behaviour. Although BASF would probably have preferred a systemic compound, the company did not discover such a compound in its synthesis programme, at least not outside of competitors' patent claims. Under the competitive pressure of ICI/Zeneca and other agrochemical companies, BASF considered the characteristics of BAS 490F good enough to meet the various demands for pesticide innovation. Satisfying behaviour is inherent in the structural setting of pesticide innovation: when regulators demand that specific criteria be met for registration, these criteria represent satisfactory performance levels. However, registration criteria are not the only criteria to be met. In order to satisfy customer demands, however ambiguously formulated, the technical and cost performance of the new product needs to be superior to competing products. Thus, the price and performance characteristics of pesticide products already in the market pose minimum performance levels that are satisfactorily dealt with if superseded, while production costs must not be excessive.

In addition to the criteria to be met for product registration, market acceptance is the second major part of the selection environment. Agrochemical companies regularly survey farmers on their perception of crop protection problems, on their use of pesticide products, and on their assessment of the efficiency and efficacy and of the pesticide products available to them in the market. Information from farmer surveys is also important in identifying the occurrence of secondary pests and in evaluating performance gaps in the product portfolios of the various agrochemical companies. New products must perform better than established products (in terms of better efficacy in pest control or lower cost to the farmer). There is, however, no direct relation between such market information and the new product development processes. The link between the two is established through the set-up of the screening tests for biological activity. Biological screenings are essentially company-specific although there is an estimated overlap of some 60–70% between the biological screenings of the various agrochemical companies. Major crops and major pests, diseases and weeds are represented in the screenings of all agrochemical companies, but specificity of the screenings stems from the choice of which cultivars of the major crops (disease resistant or susceptible) and which pest strains (level of virulence) to include. Screening set-ups are adjusted according to information gained from surveying farmers and thus represent a slowly shifting selection environment. The specific details of screening set-ups are well-protected secrets because differences in screening set-ups may result in different evaluations of the usefulness of potential chemical leads, and thus have profound effects on the innovative success of the company. In the case of fungicide development, the screening set-ups represent a general demand for new active ingredients that have broad-spectrum activity, including the less well controlled oomycete fungi. Other important characteristics that are much desired in new active ingredients, such as systemic action, no effects on non-target organisms, no human toxicity, no leaching to groundwater, are currently not part of the initial screening for biological activity.[9] Nor can other important developments in agricultural production systems be represented in the initial screening, such as compatibility with schemes for integrated pest management.

How about the linkage between 'problems' and 'solutions' in pesticide innovation? Heuristics develop over time in relation to the existing capacity for the synthesis and screening of new chemical compounds. Large numbers of compounds are produced, any of which might be a potential solution. The design of the initial screening for biological activity is a representation of perceived market needs and thus of the innovation problem. It changes over time as the perception of market needs changes. The large and increasing number of chemicals screened for biological activity makes the screening a stochastic process. It may be somewhat 'distorted' in the sense that perceived market needs are translated into selection criteria for one sort of solution: pesticide products. The screening set-up is a construct of pesticide innovators by which the complexity of agricultural practice is reduced in such a way that it can be dealt with in the heuristics. Thus, the heuristics which guide the development of potential crop protection 'solutions' is independent of the definition of crop protection 'problems', and vice versa. It is coincidental when some match is found between the characteristics of any compound screened and the characteristics looked for in the screening set-up of that moment.[10] In this sense, the innovation process resembles a 'garbage can' (Cohen *et al.* 1972). There is no cause–effect linkage between problems and potential solutions; they become connected because they appear at the same moment in time. Of course, once a link has been established, that is, once a 'lead' has been identified, an optimisation process is started in which greenhouse and field tests are indispensable means for assuring the practical efficacy and efficiency of the many compounds synthesised.

Drivers of agrochemical innovation[11]

The strobilurin case is one of many different cases that could have been studied. The focus on the identification of a new active ingredient, or group of ingredients, may have somewhat obscured other innovations, including the development of new pesticide products by novel combinations of active ingredients, new and improved formulation types, new application technologies and new packagings. Other sorts of innovations in the context of crop protection include the introduction of bio-technology, which led to market introductions of herbicide-resistant and insecticide-producing crops (Carr in this volume) and the commercialisation of natural enemies to control pest infestations in commercial crops. Agrochemical companies compete with each other by pursuing various heuristics in search of new solutions for crop protection problems that are framed in a particular way so as to fit what these companies are good at. This is both a cause and a consequence of the competitive context in which innovation processes in the agrochemical industry take place. Three aspects can be observed from the strobilurin case, but have more general relevance (Hartnell 1996): (1) extensive regulation (aimed at the protection of farmers, consumers and the natural environment), (2) protection of intellectual property rights through patent systems, and (3) fierce market competition among R&D-based agrochemical companies which regularly develop new active ingredients and pesticide products.

Table 4.1 summarises important driving forces that shape the competitive environment and the innovation challenge for the agrochemical industry. Along with agricultural reform in Western economies, farmers have fewer resources to spend on agricultural inputs, thus reducing the value of crop protection markets. Generic producers of off-patent pesticide products who compete with low-cost products are gaining market share, especially in developing countries. New products have to meet the increasingly stringent criteria in the major, Western markets for product registration. Consumers continue to be suspicious of the environmental and health effects of pesticide products. For most (if not all) crop protection problems, at least one reasonably effective solution is already on the market, implying that new pesticide products have to compete with established products on both end-user service and price. Moreover, competition from substitute technologies such as the new genetic technologies and, although quite different in origin, non-pesticide pest control technologies is increasing in specific markets.

Table 4.1   Driving forces in the agrochemical industry (den Hond *et al.* 1999).

Power shifts along the agro-food chain → marginalisation of farming
Patent expiration → competition from producers of generic pesticide products
Increased market saturation in the industrialised markets
Rising R&D costs
Reduction in producer price incentives
Herbicide-tolerant crops & seed-based protection
Consumer preferences
Liabilities
National programmes for reduction of pesticide use

Because of such forces the industry has felt the need to increase the scale of operations in order to make up for the sharply increasing R&D costs in stagnating markets. It is reported that the minimum 'critical mass' of broad-spectrum crop protection companies has mounted to annual sales of over 2 billion US dollars (Mol 1995). The top twelve companies controlled over 80% of the 1990 world-wide sales. Voss (1995) expected a further concentration to eight pesticide producers that control over 90% of the world market by the year 2000. He was not far off the mark; Agrow reports that by 2000 seven major agrochemical companies have remained, still controlling almost 80% of the 30 billion US dollar world market (Table 4.2).

Table 4.2   Turnover of major agrochemical companies (*Agrow World Crop Protection News*, 5 January, 2 March, 13 April 2001).

	2000 sales (US$ millions)	% change *vs.* 1999 (local currencies)
Syngenta	5.888	– 2.6
Monsanto	3.885	+ 8.3
Aventis	3.701	– 0.6
DuPont	2.511	– 3.1
Dow	2.271	– 0.1
Bayer	2.252	+12.6
BASF	2.228	+39.1

The industry has responded to these forces with the consolidation of its assets and research and development (R&D) capacity. Since the early 1980s, the pesticide businesses of the (petro-) chemical industry have observed major restructurings, resulting in increased concentration. Over a period of 15–20 years, the number of companies developing and marketing new active ingredients has declined from well over twenty to slightly more than a handful in the early twenty-first century. A steady focusing of the activity portfolios in the (petro-) chemical industry can be observed. Historically, those companies were highly diversified on the basis of the marketing of by-products from R&D and production. However, takeovers, mergers and divestitures of various activities resulted in a movement towards concentration and specialisation. Thus, several companies divested their agro-divisions to competitors; for example, Shell US and Union Carbide sold their pesticide activities to Du Pont and Rhône-Poulenc respectively. Other companies divested those businesses where the new biotechnology was of little help, while concentrating their R&D capacities in the 'life science' or 'bio-science' businesses. Agrochemical businesses became part of the life science businesses. ICI's demerger of its pharmaceutical and agrochemical activities to form Zeneca in the early 1990s was the first example of a series of comparable moves. It is a consequence of the complexity in managing increasingly diverse businesses (Owen & Harrison 1995), for example in terms of the difference in risk profiles of R&D into the chemical business on one hand and life science businesses on the other. Other companies merged their agro-divisions with those of competitors to form joint ventures, such as AgrEvo (Hoechst and Schering agro-activities) and DowElanco (Dow Chemical and Eli Lilly agro-activities). Interestingly, Dow has recently bought out Eli Lilly from the DowElanco joint venture to form the new Dow Agroscience Division. Monsanto and Novartis – itself the product of a complex exchange of businesses between Sandoz and Ciba-Geigy in 1996 – have followed ICI's example of separating the fine-chemicals business, including pharmaceuticals and agrochemicals, from bulk chemistry. Rhône-Poulenc and Hoechst split off Rhodia and

Celanese respectively in preparation for their merger into Aventis, effectuated in 1999. Finally, it would appear that a new round of restructurings started at the beginning of the twenty-first century. Further concentration and specialisation is about to occur, e.g. the 1999 merger between Astra and Zeneca in pharmaceuticals was followed by a merger of AstraZeneca's agrochemical activities with those of Novartis Agro to form Syngenta in 2000. This last movement could imply that the industry has started a new strategic orientation away from the life science concept (Assouline *et al.* 2000). There are additional indications. BASF plans to divest its pharmaceutical business to concentrate further on its agrochemical business. American Home Products, recently acquired by Pfizer, sold off its agrochemicals division Cyanamid to BASF in order to concentrate on pharmaceuticals. Monsanto, likely to join forces with Pharmacia & Upjohn, is reported to consider separating its pharmaceutical and agrochemical activities (*NRC Handelsblad*, 19 April 2000).

New product development and R&D are being oriented toward developing new pesticide products with broad-spectrum activity for a small range of world cash crops: wheat, rice, soybeans, corn and cotton. About 50% of the 1995 world-wide pesticide sales are used on these five crops (data from *Industrieverband Agrar* (IVA)).

The main agrochemical companies are investing in gene technologies. In the 'first wave' of agro-biotechnology the principle aim is to make crops tolerant to pesticide products. Examples are Monsanto's 'Roundup Ready' and AgrEvo's 'Liberty Link' marketing concepts of herbicide-tolerant crops, initially maize but extended to other GM crops such as soy and cotton. Crop protection may well delink from chemistry in a 'second wave'. For the coming decade it is expected that not only pest- and disease-resistant crops will be marketed, but also crops that are otherwise genetically engineered to provide additional value, e.g. through increased shelf life, or increased nutritional or health value, or by exhibiting desired characteristics for specific industrial, pharmaceutical or food processing needs. Monsanto's managing director Bernard Auxenfans says in *Cultivar* (November 1996, special issue) that the company's strategic intent is to concentrate its efforts in the areas of food, seeds and agriculture: 'We are no longer just a plant protection products company, but also a company involved in agricultural production and processing.' Monsanto's strategic intent is echoed in the recent strategic partnership between Du Pont and Pioneer Seeds.

Whereas the early industry concepts of 'related diversification' and 'life science' were essentially technology-driven because of expected or real synergies of scale and scope in production, R&D, and product registration, the more recent divorce between the pharmaceutical and agrochemical businesses is probably better characterised as stakeholder-driven. Increasing public resistance against GM technology, and continuing regulatory ambiguity concerning the conditions under which GM technology is permitted, result in increased commercial risk. Increasing resistance in international finance against the differential returns between agrochemical and pharmaceutical businesses (10–15% *versus* 20–25%, respectively) (Assouline *et al.* 2000) is threatening stock value.

Taking a broad view of these trends in the pesticide industry, it may be concluded that the industry's innovation efforts increase agricultural yields through the development and implementation of generic crop protection technologies. World markets for agricultural inputs are further standardising, resulting in increased technical uniformity of agricultural production and strategic decision making. However, consideration of some of the driving forces as mentioned in Table 4.1, including changing consumer preferences, liabilities and national pesticide reduction programmes, as well as the highly mature state of the industry, might suggest that some companies attempt to break away from the perceived industry wisdom in order to rejuvenate their business (see Baden-Fuller & Stopford 1992), even though divergent technological strategies are not without risk (Abetti 1996).

Radicality of agrochemical innovation

In addition to discussing the process and drivers of agrochemical innovation, the outcomes are interesting too, especially in the light of the industry's claim of developing innovative solutions. Earlier research (Achilladelis *et al.* 1987; Achilladelis & Antonakis 2001) suggests that agrochemical companies which introduce more radical innovations are more successful than companies introducing relatively more incremental innovations, arguably because of first-mover advantages (notably patent protection) in the subsequent optimisation. However, this research is limited by the implicit definition of what 'radical innovations' are. Achilladelis and his colleagues consider as innovations the market introduction of new pesticide products. Considering the level of radicality of these pesticide innovations, they asked industry experts to classify the items on an exhaustive list of active ingredients as 'radical', 'intermediate' or 'incremental'. The experts are likely to have used a set of criteria that relate to the opening of a new class of chemical structures, mode of action, relative product performance, and commercial success. Along these lines, the market introduction of strobilurin products by BASF and Zeneca would count as radical innovations. Another limitation is that innovations based on biotechnology, biopesticides, and new formulations and application techniques remain out of scope.

Tushman & Anderson (1986) distinguish between radical and incremental innovations by assessing the company's underlying competence base: incremental innovations are 'competence enhancing', radical innovations 'competence destroying'. The relevance of this notion is the argumentative power in explaining why so many companies lose competitive power once a radical innovation is introduced. They simply lack the ability of acquiring the new competencies that are required to compete at the new standards. Such firms are better at exploiting their acquired competence base than at exploring new competence bases (March 1991). Their 'core capability', valuable before the arrival of the radical innovation, has turned into a 'core rigidity' because of the arrival of a radical innovation (Leonard-Barton 1992). Although 'strobilurin chemistry', to take one example, differs from the chemistry in other classes of pesticides, it is difficult to maintain that in developing competencies in strobilurin chemistry, BASF or ICI have weakened, or even destroyed, the set of competencies by which they and their competitors compete in the agrochemical industry. It is unlikely that the underlying abilities in chemical synthesis or in the testing for biological activity differ very much between the various classes of pesticide products. Moreover, for most of the companies in this industry, the heuristics that are applied to identify promising chemical leads are complimentary; the choice of leads is most likely determined by considerations of efficiency and efficacy, rather than applicability. It might even be difficult to argue that the new biotechnology, which has led to the integration of pesticide and seed businesses as explored by Monsanto and AgrEvo, is competence destroying since the same fundamental techniques are applied in both the genetic modification of crops and the identification and testing of biologically active compounds and their modes of activity. The critical competence in the strobilurin case is not strobilurin chemistry, but abilities in chemical synthesis, elucidation of the metabolic interferences of lead compounds, and the set-up of screenings for biological activity. In comparison to these abilities, the diversification of agrochemical companies into application techniques or biocontrol would constitute more radical departures from their developed set of competencies. Indeed, during the 1980s and early 1990s various agrochemical companies explored the opportunity of developing business in biological pest control, but did not consider the opportunity sufficiently interesting for commercial development, mainly because of their inability (core rigidity, Leonard-Barton (1992)) to accommodate the required scale of production, shelf life, and performance consistency (den Hond *et al.* 1999).

Henderson & Clark (1990) have expanded the Tushman & Anderson (1986) dichotomy by pointing out the critical competence that exists in arranging the various components in products and services. Core concepts in how components are arranged may be reinforced ('modular' innovation) or overturned ('architectural' innovation) (Fig. 4.2).[12] Pest control technology can be seen from this perspective as a specific arrangement of a number of components, including e.g. active ingredients, formulation chemicals, application apparatuses, and seeds or cultivars, that need to be mutually adapted to provide for effective crop protection. One can speak of the pesticide innovator's competence in renewing these various components and in rearranging their relationships, but also of the farmer's competence in making productive the proposed arrangement of these compounds. Regarding the innovator, the sort of innovations that Achilladelis *et al.* (1987) consider 'radical' – strobilurin fungicides are a case in point – would be considered 'modular' in the Henderson and Clark framework. In contrast to the Tushman and Anderson framework, the application of the new biotechnology would be considered to result in 'architectural' innovation, rather than incremental innovation. The linkages between pest, crop and pesticide product are significantly changed: rather than finding formulations of active ingredients that are effective in controlling pests in specific crops, the crop is modified either to produce effective toxins as in the case of Bt toxin-producing cotton ('Bt cotton') or to withstand the toxicity of established active ingredients as in the case of the Round-up Ready and Liberty Link concepts. The question of whether the core concepts – notably the crop – are overturned or reinforced is highly similar to the question of whether the application of new biotechnology for the introduction of GM crops is competence enhancing or destroying. However, as far as the farmer is concerned, nothing has really changed with the introduction of GM crops; it could be that the farmer actually needs less specific know-how in making the various components productive, because some of the knowledge (of which pesticide product to use in order to control pests effectively) has been integrated in the crop while the timing of pesticide applications has also been made less critical.

		Core concepts are:	
		Reinforced	Overturned
The linkage between core concepts and components is:	Unchanged	Incremental	Modular
	Changed	Architectural	Radical

Fig. 4.2 Incremental, modular, architectural and radical innovations (Henderson and Clark, 1990, reproduced with permission).

Christensen (1997) argues that the incremental–radical distinction made by Tushman & Anderson and others along the lines of competencies is misplaced, because companies are very able quickly to develop new competencies if required or desired by customers. Most of the incremental or radical innovations are 'sustaining' in the sense of reinforcing the criteria by which customers assess performance. What companies are not good at is leveraging 'disruptive' technologies, i.e. technologies that redefine customer's performance characteristics. From this perspective, the introduction of new classes of pesticide products, improvements in formulation and application technologies, but also the introduction of GM crops, such as herbicide-resistant soy and corn, or Bt cotton, reinforce the performance expectations of full pest control, total eradication and no collateral damage to valuable crops. In this respect, the introduction of new concepts such as economic damage thresholds and integrated pest management or integrated crop management (IPM/ICM) constitutes a business risk to agrochemical companies – even if it may also be a crucial strategy for keeping markets – because they modify farmers' performance expectations. Rather than promising to deliver against agronomic performance criteria such as 'total control' and increased yields, these new concepts promise to deliver against economic performance criteria by turning crop protection into a cost:benefit question of damage *versus* a variety of pest control instruments: pesticide products, tillage, crop rotation and so on, that are partially substitutes and partially complements. Thus, agrochemical companies need to develop products that can be applied in IPM/ICM pest control schemes, but they have difficulties in testing whether specific products fit in with such schemes and little say in how they are actually being defined. Farmers' movements, agricultural extension services, agronomists and, increasingly, the food processing and retail industries (van der Grijp in this volume) are better positioned to develop such schemes.

Having considered agrochemical innovation from a number of theoretical perspectives, which included both the innovator and the user of innovations, it remains difficult to maintain that the innovative activity of companies in this industry is really breaking away from the past. Innovative activity is enhancing the innovator's competence (Tushman & Anderson 1986), reinforcing the linkages between core concepts and components (Henderson & Clark 1990), and sustaining the criteria by which users assess performance (Christensen 1997).

Availability of pesticide products

This section presents results from the outcomes of innovation processes in the agrochemical industry by analysing the availability of active ingredients and pesticide products[13] in the Dutch market over a 35-year period. Pragmatic reasons such as data availability have guided the selection of the Dutch market over markets that are more important in terms of absolute turnover or in terms of relative share in the five major world crops: rice, cotton, wheat, corn and soy. However, only a handful of national markets, such as the USA, Japan and France, would be classified as 'important' in such terms.

Dutch agriculture is dominated by dairy farming and meat production; almost 50% of cultivated land is in use as grassland. Arable farming concentrates on potatoes, corn and sugarbeet. Horticulture is relatively important, notably the production of flower bulbs and the greenhouse production of flowers and vegetables such as sweet peppers, tomatoes and cucumbers. Consequently, fungicides (47%) and herbicides (30%) dominate in the Dutch pesticides market when measured as kilograms of active ingredients. Fungicides are used on potato and wheat, and in horticulture; herbicides in potato, corn and sugarbeet. Soil fumigants used to be more significant, notably in potato, but their use was restricted in the late 1990s, which resulted in a reduction to less than 25% of the average use in the late 1980s. Finally, pesticide use on Dutch arable land is among the highest in Europe (13.5 kg ai ha^{-1}), but among the lowest when measured in kilograms of active ingredient per monetary value of yields per hectare (de Jong 1999).

The year 1965 was chosen as the starting point of the analysis, since in that year registration of commercially available pesticide products was effectuated. Pesticide products need to be registered before they can legally be sold in The Netherlands (*cf.* Vogelezang in this volume). Usually, registration was for a period of ten years, after which registration could be extended. Thus, analysis of all the registrations identifies which pesticide products and, consequently, which active ingredients were potentially commercially available. In the period 1965–1999, 469 active ingredients and over 3200 pesticide products received a registration. Figs 4.3 and 4.4 show, respectively, the number of active ingredients and pesticide products that were admitted for commercial trade per year.

It can be observed that the period 1965–1982 was characterised by a rapid and constant increase in the number of active ingredients (Fig. 4.3) and pesticide products (Fig. 4.4), while the number of registration holders in the Dutch market also increased steadily by an average of almost seven per year.

The period 1982–1992 is characterised as a period of maturity. The number of registered active ingredients remained constant within a narrow band width, despite a small decrease from 1988 onwards (Fig. 4.3). The increase in the number of admitted

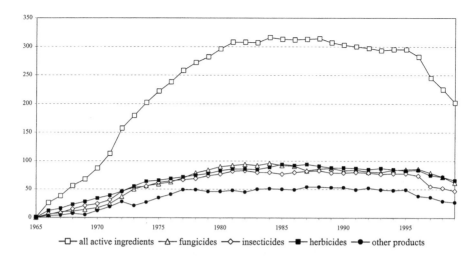

Fig. 4.3 Number of active ingredients admitted in the Dutch market, per year (den Hond 1999).

pesticide products slowed down (Fig. 4.4). This could suggest increased product differentiation and an increase in the number of derived and parallel authorisations. Further, the number of registration holders remained constant.

In the 1990s, however, a fundamental shift is observed. The number of authorised pesticide products as well as the number of registration holders decreased dramatically after 1992 (Figs 4.3 and 4.4). The number of active ingredients has decreased by one-third since 1995. Van den Bijlaard (1997) finds a similar trend in his analysis of all pesticide products in The Netherlands, including those for agricultural, veterinary, household and wood conservation uses. The most important causal factor for this shift is likely to be the introduction of an administrative levy on holding a pesticide

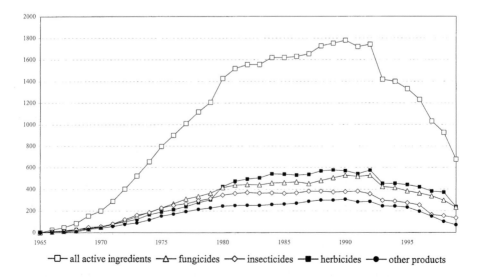

Fig. 4.4 Number of pesticide products admitted in the Dutch market, per year (den Hond 1999).

product registration (the so-called *instandhoudingheffing*). This levy resulted in a significant decrease in the number of pesticide products available in the market, because it added cost. All of a sudden registration holders needed to consider whether continuing a registration for specific products was worth the extra cost. Shortly after the introduction of the levy, the numbers of active ingredients and registration holders also diminished. This should be interpreted as a second order effect of the introduction of the levy. Attempts by the Dutch state to ban a number of notoriously risky active ingredients were ineffective because of juridical problems.

However interesting may be the dynamics of the aggregate numbers of active ingredients and pesticide products in the Dutch market, it obscures the number of introductions of new active ingredients as well as the age of the available active ingredients. It appears that, overall, the number of new active ingredients admitted in the Dutch market has remained constant at an average of nine since the mid-1980s.[14] The average during the 1960s and 1970s was around 15 per year. Thus, the number of new registrations for active ingredients has decreased, whereas the number of active ingredients disappearing from the market actually increased, especially after the introduction of the registration levy.

Fig. 4.5 details the year of introduction of the active ingredients and pesticide products that were registered at 1 January 1999. Half of the number of active ingredients registered by 1 January 1999 were introduced after 1985, whereas half of the number of registered pesticide products were registered after 1988. This would suggest a good deal of product renewal if it is considered that normally a product registration is granted for a period of ten years.

However, this suggestion needs to be qualified with pesticide usage data. When measured in kilograms, fifteen out of the twenty most applied active ingredients were introduced before 1980, and only three out of these twenty after 1990 (de Jong 1999).

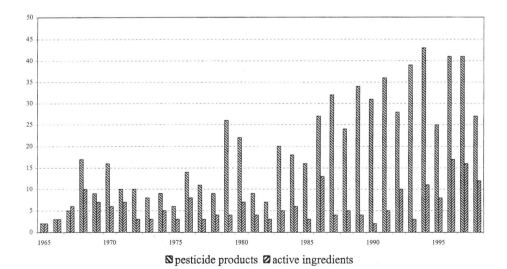

Fig. 4.5   Year of introduction of pesticide products and active ingredients registered in The Netherlands, at 1 January 1999 (den Hond 1999).

Moreover, de Snoo (1999), applying the 'environmental yardstick approach', argues that four out of these twenty contribute significantly to the environmental impact of pesticide use in The Netherlands. Even if kilograms of active ingredient is taken as a very poor indicator of environmental impact (Parris & Yukoi in this volume), and the application rates between long established and recent pesticides may differ by a factor of 1000 or more, it is safe to conclude that many farmers stick to using the older products that they know well, and that this is likely to be a significant contribution to the environmental impact of pesticide use. Indeed, if the new, low dose rate pesticides were to substitute for the long-established, high dose rate pesticides, a significant change should have been observed. However, the 25–30% reduction of pesticide use that has been observed in The Netherlands during the 1990s is to be ascribed to the curtailing of soil fumigant use.

Conclusion

This chapter attempted to shed some light on innovation in the agrochemical industry. The question to be addressed was *how* the industry is innovative.

In the agrochemical industry's technological paradigm, chemistry is believed to provide solutions for the crop protection problem, as well as for the problems of pest resistance and intra-industry competition. However, there does not appear to be a direct connection between the discovery of new chemical compounds that offer leads for pesticide development and the problems that such compounds are supposed to solve. They get connected in the screening for biological activity. At the time of screening, those compounds are selected whose characteristics, which are not known beforehand, match the characteristics looked for as represented by the set-up of the screening.

As the strobilurin case suggests, at the detailed level of science-based development work, industrial researchers adopt the technical logic of the natural sciences. They collect similar empirical evidence and formulate and test highly similar hypotheses. The lack of direct communication among researchers at various companies does not prevent them from developing similar heuristics and organisational routines. Supposedly, factors such as professional formation and culture structure the localised social processes that underlie heuristics and organisational routines.

The results of the innovation process depend on how the innovator deals with the various, often conflicting, constraining factors in the selection environment. Contingent factors such as critical (research) events compel actors to decide on whether and how to continue research. There is a 'luck' factor at hand which influences whether a company is successful in pesticide innovation. However, companies do follow deliberate strategies to increase their commercial success.

In a mature market, differentiation from competitors, e.g. by new product development, is considered to be an appropriate business strategy. Indeed, the agrochemical industry is very innovative in the sense that it continues to develop and market new pesticide products based on new active ingredients. It is not unlikely that ever new product groups will be introduced. One could argue that such introductions are 'radical' innovations (Achilladelis *et al.* 1987), because they exhibit new modes of biological action, thus reducing the risk of pests building up resistance, and constitute new chemistry. However, by detailing more precisely what is 'radical' a more differentiated picture appears. In the Tushman & Anderson (1986) framework, most of the industry's innovations are 'competence enhancing'; it would even be difficult to argue that the underlying competence in developing GM crops is 'competence destroying' to this industry. From Henderson & Clark's (1990) perspective, the introduction of new active ingredients and pesticide products is 'modular' in the sense that they complement or substitute products already in the market without modifying the underlying principles of crop protection in the farm's production system. GM technology would be 'architectural' to the innovator since it establishes new links between the components, but 'modular' to the farmer since from his perspective it leaves unaltered the links between crops and pests. Moreover, in Christensen's framework, it sustains the performance criteria that farmers are used to applying. By contrast, biological control and related technologies would be competence destroying, overturning the linkages between the various components, and disruptive to farmers' performance criteria. Although companies in this industry have explored in this area, they have not been able to make biological control a commercial success, which is partly due to core rigidities in their size, production and organisation principles.

Finally, the longitudinal analysis of product availability in the Dutch market suggests that, on the one hand, new products are introduced in the market quickly. New products are available to farmers at a relatively short delay. The speed at which industry manages to introduce new products is dependent on the efficacy of registration procedures. On the other hand, and despite the withdrawal of significant numbers of pesticide products and consequently of active ingredients from the Dutch market, more than half of the number of pesticide products available are older than 15 years, among which are the most used and most environmentally risky products. It would appear that the industry's innovations are relatively quickly introduced in the market, indeed at a quickening rate, although market penetration appears to be much slower given the rather stable pesticide use data in kilograms of active ingredient. A number of farmers apparently stick to using the old products they know well.

Notes

1 http://www.corp.aventis.com/cropsc/position/position.htm, 23 February 2001.
2 http://www.monsanto.com/monsanto/about_us/business_purpose_mission/default.htm, 23 February 2001.
3 http://www.syngenta.com/en/syngenta/index.asp, 29 May 2001.
4 The empirical data and analysis on strobilurin fungicides have been condensed from den Hond (1998a). References have been omitted, except in cases of direct quotations, in order to enhance readability.
5 The toxophore is that part of the active ingredient that is responsible for biological activity by binding to the receptor in the target cell and hence inhibiting crucial biochemical

processes, in this particular case the inhibition of mitochondrial respiration. Other parts of the active ingredients moderate or enhance its efficacy.

6 BAS 490F was the first active ingredient to be submitted under the new EU registration directive 91/414 whereas ICI A5504 was the last compound to be registered under the old system before the 91/414 Directive.

7 It would be somewhat beside the point to relate the emergence of this particular trajectory and those of other systemic fungicides around 1980 to a 'technological imbalance' (Rosenberg 1969) of the resistance problem. On the one hand, resistance is a consequence of the large-scale agricultural use of pesticides, not the sort of 'internal compulsions and pressures [in a complex technology] which […] initiate exploratory activity in particular directions' that Rosenberg (1969) seems to point at. On the other hand, neither was the resistance problem the reason to start exploring the potential of naturally occurring compounds, nor were solutions to the resistance problem solely sought with the natural compounds' heuristics; other recent systemic fungicides have been discovered through other heuristics.

8 In a generic way, of course, industry does take issue with many established norms. A well-known example is the EU residue level of active ingredients in (ground)water for drinking purposes. A further discussion on the role of science in the regulatory process is by Irwin & Rothstein in this volume.

9 However, efforts are being made to predict such environmental behaviour of chemical structures based on molecular modelling of structure–activity relationships. If these attempts are successful, they may have significant impacts on the agrochemical innovation processes, because it would be possible to eliminate chemical structures from the screening process because of unfavourable toxicity and environmental impacts even before the structures are synthesised (Magee 1995).

10 An illustration to this point is the development of compounds that induce systemic resistance in plants. Ciba-Geigy started screening for such compounds in 1980, but had to develop a new screening protocol. Several compounds that had been screened before passed the new test, and indeed a compound that had already been screened in the early 1970s was positive on the new test. However, this compound did not become a commercial product (den Hond 1998b).

11 This section is an update and extension from den Hond *et al.* (1999).

12 'From the point of view of a battery maker, a car battery with a much longer life might be a radical innovation. For a car maker, it would be an example of modular innovation if the new battery were used in the traditional way, since the rest of the car's design could be unaffected. On the other hand, a battery-powered car, if it used well established components but represented a major improvement in performance, would be an architectural innovation' (Fairthlough 1994).

13 The empirical data and analysis in this section have been condensed from den Hond (1999).

14 A peak of 21 new active ingredient registrations in 1992 is incidental. It can be ascribed to an administrative decision considering a number of applications that the registration authorities had not been able to decide upon because of internal conflict. In a quick procedure many of these applications were registered (den Hond 1999).

References

Abetti, P.A. (1996) The impact of convergent and divergent technological and market strategies on core competencies and core rigidities: An exploratory study. *International Journal of Technology Management*, 11 (3/4), 412–24.

Achilladelis, B. & Antonakis, N. (2001) The dynamics of technological innovation: The case of the pharmaceutical industry. *Research Policy*, 30, 535–88.

Achilladelis, B., Schwarzkopf, A. & Cines, M. (1987) A study of innovation in the pesticide industry: Analysis of the innovation record of an industrial sector. *Research Policy*, 16 (2), 175–212.

Assouline, G., Joly, P-B. & Lemarié, S. (2000) Biotechnologie végétale et restructurations de l'agrofourniture: Un horizon stratégique marqué de fortes incertitudes. *Economie et Société*, (July).

Baden-Fuller, Ch. & Stopford, J.M. (1992) *Rejuvenating Mature Business: The Competitive Challenge*. Routledge, London.

Beautement, K., Clough, J.M., de Fraine, P.J. & Godfrey, C.R.A. (1991) Fungicidal β-methoxyacrylates. *Pesticide Science*, 31, 499–519.

van den Bijlaard, M.F. (1997) Ontwikkelingen in het aantal toegelaten bestrijdingsmiddelen, 1990–1995. *Kwartaalberichten Milieustatistieken (CBS)*, 97/1, 25–9.

Brent, K.J., (1995) *Fungicide Resistance in Crop Pathogens*. GIFAP, Brussels.

Cohen, M.D., March, J.G. & Olsen, J.P. (1972) A garbage can model of organisational choice. *Administrative Science Quarterly*, 17, 1–25.

Christensen, C.M. (1997) *The Innovator's Dilemma*. Harvard Business School Press, Boston.

Dosi, G. (1982) Technological paradigms and technological trajectories. *Research Policy*, 11, 147–62.

Evans, D.A. & Lawson, K.R. (1992) Crop protection chemicals: Research and development perspectives and opportunities. *Pesticide Outlook*, 3 (2), 10–17.

Fairthlough, G. (1994) Innovation and Organization. In: *The Handbook of Industrial Innovation* (eds M. Dodgson & R. Rothwell), pp. 325–36. Edgar Elgar, Cheltenham.

Hartnell, G. (1996) The innovation of agrochemicals. *Research Policy*, 25, 379–5.

Heaney, S., Slawson, D., Hollomon, D.W., Smith, M., Russel, P.E. & Parry, D.W. (1994) *Fungicide resistance*. BCPC Monograph No.60. BCPC, Farnham.

Henderson, R.M. & Clark, K.B. (1990) Architectural innovation: The reconfiguration of existing product technologies and failure of established firms. *Administrative Science Quarterly*, 35, 9–30.

den Hond, F. (1998a) On the structuring of variation in innovation processes: A case of new product development in the crop protection industry. *Research Policy*, 27 (4), 349–67.

den Hond, F. (1998b) Systemic acquired resistance: A case of innovation in crop protection. *Pesticide Outlook*, 9 (2), 18–23.

den Hond, F. (1999) Toegelaten bestrijdingsmiddelen in Nederland sinds 1965. In: *Bestrijdingsmiddelen en milieu* (eds G.R. de Snoo & F.M.W. de Jong), pp. 19–35. Jan van Arkel, Utrecht.

den Hond, F., Groenewegen, P. & Vorley, W.T. (1999) Globalization of pesticide innovation and the locality of sustainable agriculture. *American Journal of Alternative Agriculture*, 14 (2), 50–58.

de Jong, F.M.W. (1999) Hoeveelheid gebruikte bestrijdingsmiddelen. In: *Bestrijdingsmiddelen en milieu* (eds G.R. de Snoo & F.M.W. de Jong), pp. 37–51. Jan van Arkel, Utrecht.

Leonard-Barton, D. (1992) Core capabilities and core rigidities: A paradox in managing new product development. *Strategic Management Journal*, 13, 111–25.

Magee, P.S. (1995) Searching for clues to mechanism and new directions in synthesis through SAR and modelling. In: *Eighth International Congress of Pesticide Chemistry: Options 2000* (eds N.N. Ragsdale, P.C. Kearney and J.R. Plimmer), pp. 34–41. American Chemical Society, Washington, DC.

March, J. (1991) Exploration and exploitation in organizational learning. *Organization Science*, 2 (1), 71–87.

Mol, A.P.J. (1995) *The Refinement of Production: Ecological Modernization Theory and the Chemical Industry*. Van Arkel, Utrecht.

Nelson, R.R. & Winter, S.G. (1982) *An Evolutionary Theory of Economic Change*. Belknap Press, Cambridge (Mass).

Owen, G. & Harrison, T. (1995) Why ICI chose to demerge. *Harvard Business Review*, 73 (2), 133–42.

Paulino, S.R. (1997) *Réglémentation environnementale et processes d'innovation dans le secteur des phytosanitaires.* PhD thesis, University of Toulouse.

Rogers, E.M. (1995) *Diffusion of Innovations*. The Free Press, New York.

Rosenberg, N. (1969) The direction of technological change: Inducement mechanisms and focusing devices. *Economic Development and Cultural Change*, 18, 1–24.

Sauter, H., Ammermann, E. & Benoit, R. *et al.* (1995) Mitochondrial respiration as a target for antifungals. In: *Antifungal agents* (eds G.K. Dixon, L.G. Copping & D.W. Hollomon), pp.173–191. BIOS Scientific Publishers, Oxford.

Schwinn, F. & Geissbühler, H. (1986) Towards a more rational approach to fungicide design. *Crop Protection*, 5, 33–40.

Spender, J.C. (1980) *Industry Recipes*. Basil Blackwell, Oxford.

de Snoo, G.R. (1999) Milieubelasting van bestrijdingsmiddelen. In: *Bestrijdingsmiddelen en milieu* (eds G.R. de Snoo & F.M.W. de Jong), pp. 71–94. Jan van Arkel, Utrecht.

Stetter, J. (1993) Trends in the future development of pest and weed control: An industrial point of view. *Regulatory Toxicology and Pharmacology*, 17, 346–70.

Tushman, M.L. & Anderson, P.C. (1986) Technological discontinuities and organizational environments. *Administrative Science Quarterly*, 31, 439–65.

Vergragt, P.J. (1988) The social shaping of industrial innovations. *Social Studies of Science*, 18, 483–513.

Voss, G. (1995) Crop protection markets and technologies towards 2000: Opportunities and constraints. *Gewasbescherming*, 26 (5/6), 177–8.

Chapter 5
Regulatory Science in an International Regime
An Institutional Analysis

Alan Irwin and Henry Rothstein

Introduction

According to the Royal Commission on Environmental Pollution in its 21st Report (RCEP 1998:9):

> Environmental regulation has become more and more dependent on the advice of scientists ... . Yet the changed character of environmental concerns has highlighted the extent to which there are uncertainties in scientific assessments, and the scope for different perceptions of the issues involved. In some cases the interpretations and reassurances originally offered by governments have been shown to be mistaken ... . This has eroded trust in environmental regulation, which has also been undermined by the scope for evidence to be interpreted in different ways.

As the Royal Commission has observed, science is the 'essential basis' for environmental standards. However, rather than providing one single and definitive solution to environmental questions, science today often offers uncertainty, conflicting interpretation and partial evidence. Furthermore, and as the Royal Commission suggests, the public handling of scientific uncertainty and disagreement with regard to environmental regulation has caused difficulties for public trust and confidence.

The point is not to undermine the importance of scientific evidence for environmental control. Instead, the implication is that a cautious approach is required which recognises the limitations and uncertainties associated with all data sources. When it comes to scientific research at the interface between industrial innovation and government regulation – what we will term 'regulatory science' – it is important to consider what counts as 'sufficient evidence' and the precise manner in which technical investigations are to be evaluated. In situations of technical uncertainty and inevitable complexity, the very framing of scientific questions – in other words, what gets asked and how – becomes crucial to environmental decision making. At this point, we must recognise that these issues are not exclusively 'scientific' in character. Judgements as to the required burden of proof, the balancing of risks and benefits, and the requirement for external scrutiny are inevitably social and political, as well as technical, in character.

For the Royal Commission, it follows that transparency is all-important so that these assumptions, limitations and uncertainties can be clearly identified when decisions are being taken. However, and as we will argue in this paper, such a general conclusion needs to take account of the particular contexts within which regulatory science actually operates. In other words, before we can reach any recommendations for how the relationship between science and regulatory policy might be improved, it is necessary to examine in more empirical detail how this relationship currently works in practice. This is all the more important given the fact that 'regulatory science' is not primarily pure or 'curiosity-driven' but an area where commercial concerns and externally set deadlines are of necessity very much to the fore. What happens to the character of science when it is conducted within such an economically and institutionally loaded setting?

In this short chapter, we will consider the contemporary character of regulatory science in one specific area: pesticide regulation. Our aim will be to describe industrial, governmental and scientific activities in this area and to draw out some of their key characteristics. As we hope to suggest, regulatory science and compliance testing are not just 'applied science'. Instead, this is a field with its own technical and institutional structures which are partly 'scientific' in character but which also overlap with the changing social, political and economic climate for pesticide development. These structures in turn have important consequences for the practice of industrial innovation, environmental protection and regulatory compliance testing. Having considered the contemporary character of regulatory science, we will conclude by briefly assessing the implications for public policy: including those for current calls towards greater transparency in the regulatory process.

Pesticide regulation in the UK

Statutory pesticide regulation in the UK dates back to the mid-1980s. Before that, a voluntary scheme operated on the principle that only pesticides approved by government experts would be supplied. This voluntary scheme was introduced in 1957 in response to growing general concern and, specifically, the deaths of seven agricultural workers from pesticide poisoning in the late 1940s (Gillespie 1979). Under the voluntary arrangements, pesticide notification was a low priority for companies and was typically the responsibility of ex-field trials officers with relatively limited promotional prospects. The community of pesticide professionals was small and approval was often negotiated through informal contacts between industry and government officials (Irwin *et al.* 1997). Whilst companies could afford to take a relaxed attitude towards the UK system since the UK market was small, they nevertheless had to pay more attention to the United States' regulatory system which dominated the international market (*cf.* Vogelezang-Stoute in this volume). This sometimes had the consequence that companies carried out more tests than the required minimum. Meanwhile, British officials were reluctant to demand more stringent tests because they risked companies withdrawing from the UK market (Tait 1976; Gillespie 1979). Equally, such moves risked a collapse of the voluntary scheme if companies subsequently chose to withdraw.

Corporate strategies in relation to regulation were therefore relatively relaxed, but a number of pressures were building on agrochemical companies. The first was the decreasing pace of innovation within the agrochemical industry. From the 1950s to the 1970s pesticide innovation proceeded briskly, but from the mid-1970s the innovation rate started to decline as the industry found it increasingly difficult to find new pesticides that could compete against a wide range of already existing cheap products (Tait *et al.* 1991). The costs of innovating a new 'active ingredient' doubled in just ten years to around \$40m by 1990 with as many as 40 000 candidate compounds being screened for each active ingredient actually developed (Engel *et al.* 1990). Indeed, innovation costs for agrochemical companies have been such that the sector has seen wide-scale mergers and acquisitions in recent years (*cf.* den Hond in this volume).

An increasingly strict regulatory environment presented a second and related pressure on innovation. Regulatory regimes, such as the EU's low limit for pesticide residues in drinking water, have undoubtedly presented problems for agribusiness. At the same time, others have argued that increasing regulatory demands have stimulated as much as inhibited innovation (Tait *et al.* 1991). Certainly, agrochemical companies have not always resisted regulation. The interests of agrochemical companies converged with environmental groups, for example, when the UK introduced statutory regulation in 1986 to meet European demands. Agrochemical companies in that instance welcomed statutory regulation as a way of eliminating 'cowboys', creating barriers to market entry and providing a platform to help shape the development of the European regulatory system for pesticide registration (introduced in 1991).

A third pressure has emanated from the public and pressure groups. The implication of organochlorine pesticides in the declining population of many wildlife species sparked the emergence of an anti-pesticide environmental movement in the 1960s. Public attitudes as expressed in a range of surveys still suggest high concern over pesticide risks. For example, a small majority of European consumers view the 'total absence of pesticides' as a key requirement for safe food (INRA 1998), and a third of UK consumers identify pesticides as an 'extremely' or 'very' dangerous threat to themselves and their family (SCPR 1998).

While there is little evidence that public pressure has made any direct impact on corporate strategy (Tait *et al.* 1991: 53), many in agribusiness would acknowledge an indirect pressure via the regulatory regime and upstream commercial intermediaries. For example, the EU's Drinking Water Directive limit was set in 1980 in the context of considerable environmental concern, but at the time, agribusiness did not contest the limit, unaware that it would present a problem. Only later, when it was discovered that some chemicals were contaminating drinking water far in excess of permitted levels, did business contest the regulations (albeit unsuccessfully). When the European Commission finally proposed to revise the limit in the late 1990s, Greenpeace orchestrated a mass write-in campaign swamping the Commission with 12 000 letters of protest, making the proposal politically unsaleable. The limit was then maintained (Hood *et al.* 2001). At the same time, agrochemical companies have come under indirect pressure from large retailers and food processors who – themselves sensitive to public demands and pressure group activities – effectively operate their own secondary regulation on the use of pesticides by farmers.

Increasingly, therefore, agrochemical companies have been forced to pay attention to the changing commercial, regulatory and political climate. In particular, a number of costly regulatory failures have forced the industry to wake up to the need for better business risk analysis and regulatory responsiveness. In the rest of this chapter we consider how agrochemical companies and regulators have met these challenges. We focus, in particular, on the impacts this changing context has had on the innovation process and on the character and practice of regulatory development and compliance. We then go on to consider the implications of these developments in the light of contemporary thinking about science, the public and technological innovation.

Regulatory science in action

The new commercial, regulatory and political context outlined above has forced agrochemical companies to change their regulatory compliance and innovation strategies. The days are long gone when regulatory compliance was tacked onto the end of the innovation process. Agrochemical companies now have departments dedicated to regulatory affairs staffed by multidisciplinary teams of personnel trained in science, law and regulatory politics. These teams are involved in the innovation process from the beginning of the search for new active ingredients through to regulatory approval, ensuring that commercial, regulatory and political risks are anticipated and handled adequately at all stages. This 'process orientated' innovation strategy has created a new breed of risk managers within corporations, enhancing the responsibilities of those who in earlier generations would have had a more circumscribed technical role. Such development is not unique to agrochemicals – nor to chemical risk – and is found in other commercial sectors. Indeed, this development parallels more general trends in corporate governance epitomised by the 1999 Turnbull ICAEW report which set out risk analysis and risk management as a central feature of good corporate management (Hutter & Power 2000).

Given this reorganisation of innovation and product compliance strategies within corporations and the regulatory changes within Europe, what have been the consequences for scientific research as conducted both by companies and the state? Previous commentators have characterised the meeting of science and regulation (or as we have termed it 'regulatory science') as a hybrid domain intermixing often uncertain science with highly politicised environments (Jasanoff 1990; Shackley & Wynne 1995; Irwin *et al.* 1997). In the rest of this section we describe the institutional organisation and dynamics of regulatory scientific activities in relation to agrochemicals.

The plant protection products registration Directive (CEC 1991) represented an attempt at standardising approval practices across Europe. The registration process is complex and two-tiered. Active ingredients are registered at the EU level (using a rapporteur system) whilst products containing the active ingredients are registered at the member state level according to common standards (subject to regional ecological variation) and member states are expected mutually to recognise each other's approvals. This new regime is advantageous to agrochemical companies in so far as less repetitious testing is required to obtain approval for individual national markets. There is also much greater transparency and consistency across national regulatory systems. At the same time, the increase in the range of testing and the need for member states to adapt their systems to the new regime poses challenges both for business and national regulatory systems (*cf.* Vogelezang-Stoute in this volume).

For R&D active companies, regulatory scientific activities now start at the beginning of the innovation process. Thus, potentially efficacious but unregisterable compounds are screened out by routinely putting chemicals through a standard series of regulatory compliance tests at well-defined stages of their development (*cf.* Govers *et al.* in this volume). Although these standards and tests are well defined, the sheer scale of work required to get a new active ingredient registered – involving roughly three-quarters of a tonne of documentation – means that a prime value is placed on confidence in the process. This mirrors what has been referred to in other contexts as an 'audit explosion' where paper trails are taken as a key indicator of compliance in the management of complex systems (Power 1997). Control over testing is therefore critical for an agrochemical company and a high premium is placed on quality and security in the selection of a laboratory (whether in-house or contracted out). Laboratories may be chosen because they are local and overview of their work can be a relatively simple matter, or because company representatives know their personnel or quality of work or both. Still other laboratories may be used because they are staffed with ex-regulatory personnel who have a thorough understanding of the regulatory implications of the work and maintain good contacts with current regulatory officials.

The significance of relations between scientific, commercial and regulatory staff should not be underestimated. In the UK at least, the pesticides world is small so that key players know each other by name and there is a perceived unity of purpose. Good relations between commercial, scientific and regulatory staff offer opportunities to resolve efficiently questions that arise during the sifting and review of a large amount of information. While summaries of dossiers pass through further stages of scientific review, these early stages are crucial in framing the presentation of information.

Commercial regulatory scientific activities are not confined, however, to regulatory compliance testing. The agrochemical industry has also been able to 'feed forward' into regulation by active involvement in the development of the European regime, primarily through the European Crop Protection Association expert committees. The agrochemical industry is the repository of much knowledge about the action of pesticides in the environment, and has used the creation of a new European regime as an opportunity to take a lead in developing testing protocols (Rothstein *et al.* 1999). In this way, business has had an opportunity at least to counter-balance the agenda-setting role of regulators.

From a UK perspective, the pesticides regulatory agency – the Pesticides Safety Directorate (PSD) – has expanded its scientific capacities significantly since the introduction of statutory pesticide regulation in the UK in 1986. This has been, in part, due to pressure from the agrochemical industry demanding more efficient processing of registration applications. As already noted, this has positioned the UK to play a significant role in the development of the European regime and, as a consequence, the PSD is viewed as one of the 'senior partners' amongst member states in this field.

Whilst the PSD takes responsibility for processing registration applications, recommendations on registration are made by an independent advisory body: the Advisory Committee on Pesticides (ACP). Given inevitable time constraints, this system relies on considerable trust between the PSD and ACP. The perception of agrochemical company representatives we have interviewed is that the working relationship between PSD and the ACP has improved over the last ten years, with the PSD often able to anticipate how the ACP will react, though it is, in the words of one business representative, 'never cut and dried'. At the same time, there is still a risk of regulatory capture within the system. In contrast to a number of Scandinavian and US regimes, the PSD sees itself very much as a 'hands-on' agency, viewing its relationship with industry as co-operative and positive in character: as 'two sides of the same coin', according to one senior civil servant. Officials we spoke to from ministries concerned with Health and the Environment considered that they had an important role in balancing the risk of capture.

By the same token, the existence of a close established institutional network means that organisations less well connected to the agency, in particular little-known (within the UK) foreign laboratories, may have greater trouble negotiating and meeting regulatory requirements than UK based or well-known organisations. Furthermore, at the European level, agrochemical company representatives suggested that there were a number of strategic issues that needed to be taken into account when trying to obtain European-wide approval. These issues include the perceived efficiency of the agency reviewing the submission, its reputation and the likelihood that the member state with the biggest market would accept that agency's evaluation.

The expansion of regulatory scientific activities in the UK has been paralleled in a few member states, but this has predominantly been in northern Europe, with southern member states generally deferring to the north. Significantly also, this expansion has not been mirrored at the level of the European Commission. Responsibility for pesticides is distributed across several Directorates-General with only a handful of officials – e.g. in Agriculture, Environment and Health, and Consumer Protection – dealing directly with pesticide regulation. Given the scale of the processes of approval and the programme of review of old pesticides, this contrasts dramatically with arrangements in the USA. Instead, the system relies on the work of member states, and this has provided the opportunity for the PSD to take a leading role in the EU. At the same time, there are political dimensions to the way that pesticides are handled by the Commission, and different strategies are employed to negotiate those politics. As one senior official put it, 'Meetings of scientists tend to take place outside Brussels. If we have them in York, the political types don't want to turn up to a boring technical meeting, so the meetings are more scientific.'

Whilst compliance testing and safety reviews are one site for observing the social shaping of regulatory design and practice, there is an important and related dimension to regulation in this field. There is a striking consensus amongst regulators, industrialists and a number of 'green' NGOs that many problems of the regulatory system lie not in the process of safety reviews (which were generally regarded, at least from the UK perspective, as high quality) but in the processes of implementation and enforcement. Problems of intentional or unintentional misuse by farmers, illegal import of banned pesticides from outside the EU, and poor or badly organised enforcement systems by member states were cited as serious difficulties with which the regulatory system was failing to cope.

In part, some of these problems can be seen as a consequence of the way in which regulation is handled at the point of standard setting. A reluctance to ban, but with the imposition of ever tighter restrictions on usage as a way of appeasing business, inevitably puts more pressure on the enforcement of a regulatory regime. Yet enforcement is not only highly complex and expensive but also a politically sensitive issue for the EU, as witnessed in the high profile debates about enforcement of anti-BSE rules in member states. In the pesticides case, the European Commission has focused on 'doing the do-able' (standard-setting) while leaving the hard implementation issues to individual member states.

In part consequence, there have been novel developments in the management of such pressures on pesticide policy. In the UK, for example, the PSD has established a Pesticides Forum which brings together stakeholder groups in an attempt to manage some of the differing and conflicting demands on the implementation of policy and development of pesticide strategy. Demands for openness have been met with records of pesticide contamination in drinking water made public (a move echoed recently at the EU level) and a limited 'name and shame' policy for identifying retailers of food contaminated with unacceptable levels of pesticide residues. Such moves, however, have been slow in implementation, and are generally orientated towards the most obviously 'public' aspects of pesticides policy. At the same time, they have not engaged in what are seen as the 'technical' dimensions of standard setting.

Discussion

This chapter began by describing some of the challenges to scientific evidence within environmental regulation. Subsequent sections have emphasised that a series of enmeshed and cross-cutting interconnections currently operate in this dynamic and responsive area. In a relatively short period of time, a national and international network of scientists, industrialists, contract laboratories and government institutions has become firmly established.

Summarising some of the main characteristics of regulatory science, a number of general points emerge:

(1) Most obviously, our discussion presents a political economy of regulatory science that seeks to go beyond traditional explanations of science as driven by the quest for knowledge. In particular, we identify European integration and pressures from large non-EU markets as the prime engines of change within EU regulatory science institutions.

(2) We have also highlighted the dominantly private institutional character of regulatory science, and the significance of innovatory and regulatory pressures

driving that science forwards. A related, and surprising, finding is the relatively minor role of the public sector beyond standard-setting and approvals, and the significance of the enormous resource pressures on even those activities: particularly at the European level. Whilst this is not necessarily a negative characteristic, it does raise issues of the best balance between the public and private sectors in the sponsorship of regulatory science.

(3)   More broadly, our discussion indicates the changing character of contemporary science in one important area of practice. In such a situation, it becomes reasonable to enquire what constitutes good science, and how this definition is changing as a consequence of the factors identified by bodies such as the RCEP.

There are a number of possible responses to this characterisation of regulatory science. One could conclude that regulatory science and compliance testing are less about science itself than the social and political organisation of risk management strategies. Seen in this way, pesticide innovation and development is primarily an economic and institutional process with science playing only a secondary role. Put like this, the challenges of developing safer agrochemicals are not essentially 'scientific' in character but instead concern wider social debates over the future of agriculture and the reliance society places on institutions to manage risks from what are, after all, 'designer toxic chemicals'.

A related response would be to argue that the 'scientific' and the 'economic' or 'ethical' dimensions of product development and usage may occasionally overlap but should as a matter of principle be kept apart. The Royal Commission on Environmental Pollution expressed this viewpoint: 'A clear dividing line should be drawn between analysis of scientific evidence and consideration of ethical and social issues which are outside the scope of a scientific assessment' (RCEP 1998). From this perspective, a solid 'fire wall' should be built between 'scientific' and 'social' factors. Certainly, British industrialists in this field are inclined to argue that science should not be corrupted by 'political' considerations. In practice, this suggestion often precedes the accusation that, whilst a sagacious mix of science and common sense rules in the UK, other European countries are inclined to transgress the science–politics distinction.

The previous discussion indicates that such a fire wall is very difficult to maintain in this area. Rather than 'science' and 'society' being kept at a distance from one another, we have instead suggested that regulatory science offers many examples of their inter-connectedness and, indeed, inseparability. Thus, the technical rationale for a particular test is bound to draw upon a sense of its political as well as scientific legitimacy. Decisions as to the degree of caution that should be exercised when, for example, assessing the assimilative capacity of a particular environment – or the treatment of improbable, but not unimaginable, events – inevitably depend upon social as well as scientific judgement. From our perspective, this hybridity of decisions does not represent a weakness – nor a criticism of those involved (as if with greater effort they could be properly scientific) – but instead suggests the inevitable character of judgements in this area.

The general conclusion from this discussion is that we are no longer dealing with an old-fashioned linear process within which the 'technical' can be sorted out in advance of the 'social/ethical' (as some risk assessment models still imply). Instead, we are discussing a more complex world where science (and scientists) cannot be restricted to the laboratory and where non-scientists cannot remove themselves from the framing of technical questions and the assessment of data. Thus, the whole thrust of the 'process orientation' described above is that the former neat technical–non-technical distinctions are at the least beginning to blur. It is important for us to stress that scientific expertise is clearly essential in designing tests and evaluating outcomes – a non-specialist can hardly give an informed opinion about the minutiae of genotoxic carcinogenicity or toxicokinetic experimentation. However, questions of what constitutes safety, or acceptability, are intimately related to the character of socially created risk management systems. What goes on in the laboratory cannot easily be divorced from the shaping forces of a complex international regulatory system which influences technological innovation in the agrochemical firm and determines how implementation works in the field and in the local marketplace.

It follows that regulatory science is not simply a matter of technical application but a more subtle blend of judgement, cross-disciplinary expertise and response to a range of contextual pressures. In this, our account of regulatory science also represents a critical response to the call for greater transparency and public engagement which has recently emanated from bodies such as the Royal Commission, the House of Lords Select Committee on Science and Technology (SCST 2000) and the UK Department of Trade and Industry (DTI 2000). Greater openness and consultation in this area does indeed seem a pre-requisite to increased levels of public trust and confidence. As we have noted, there have been some slow and cautious moves towards this in some of the more 'public' aspects of the regulatory process, albeit restricted to stakeholder consultation over policy concerns.

However, certain characteristics of regulatory science as described in the previous section – including its mix of science and economics, public and private, technical and cultural elements – make the rationale for judgements difficult to communicate since they are often based upon implicit understandings of what will be acceptable in scientific, economic and political terms. The consequent challenge is to transport such tacit and private understandings into a more public arena where very different experiences and areas of expertise may be brought to bear.

Of course, the impracticability of building a fire wall around science increases the complexity of this operation since public debate cannot be restricted simply to the 'non-scientific' dimensions of pesticide safety. It might therefore appear entirely reasonable that a broader range of voices should be heard within decision making processes. Nevertheless, and given what we have seen of the actual operation of regulatory science in this area, it would be foolish to suggest that a meaningful policy shift towards greater transparency is straightforward or without risks. Such a decisive shift in the public engagement with environmental standards will require more than sweeping assertions of the need for participation, democracy and dialogue. It will also require a practically informed and sociological understanding of the contemporary character of regulatory science and its relationship to the innovation process.

References

CEC [Council of the European Communities] (1991) Council Directive of 15 July 1991 concerning the placing of plant protection products on the market (91/414/EEC). *Official Journal of the European Communities*, No. L230.

DTI [Department of Trade and Industry] (2000) *Excellence and Opportunity: A Science and Innovation Policy for the 21st Century*. HMSO, London.

Engel, J., Harnish, W. & Staetz, C. (1990) Challenges: The industrial viewpoint. In: *Safer Insecticides:Development and Use* (eds E. Hodgson & R. Kuhr), pp 551–72. Dekker, New York.

Gillespie, B. (1979) British 'Safety Policy' and Pesticides. In: *Directing Technology: Policies for Promotion and Control* (eds R. Johnston & P. Gummett), pp. 202–24. Croom Helm, London.

Hood, C., Rothstein, H. & Baldwin, R. (2001) *The Government of Risk: Understanding Risk Regulation Regimes*. Oxford University Press, Oxford.

Hutter, B. & Power, M. (2000) *Risk Management and Business Regulation*. Centre for Analysis of Risk and Regulation, London School of Economics, London.

ICAEW [Institute for Chartered Accountants in England and Wales] (1999) *Internal Control: Guidance for Directors on the Combined Code*. ICAEW, London.

INRA [International Research Associates] (1998) La securité des produits alimentaires (Report 120). *Eurobarometer* 49. INRA, Brussels.

Irwin, A., Rothstein, H., Yearley, S. & McCarthy, E. (1997) Regulatory science: Towards a sociological framework. *Futures*, 29 (1), 17–31.

Jasanoff, S. (1990) *The Fifth Branch: Science Advisers as Policymakers*. Harvard University Press, Cambridge, MA.

Power, M. (1997) *The Audit Society: Rituals of Verification*. Oxford University Press, Oxford.

Rothstein, H., Irwin, A., Yearley, S. & McCarthy, E. (1999) Regulatory science, Europeanisation and the control of agrochemicals. *Science, Technology and Human Values*, 27 (2), 241–64.

RCEP [Royal Commission on Environmental Pollution] (1998) *Setting Environmental Standards: 21st Report*. HMSO, London.

SCPR [Social and Community Planning Research] (1998) *British and European Social Attitudes: The 15th Report. How Britain Differs*. Ashgate, Aldershot.

SCST [House of Lords Select Committee on Science and Technology] (2000) *3rd Report: Science and Society*. HMSO, London.

Shackley, S. & Wynne, B. (1995) Global climate change: The mutual construction of an emergent science-policy domain. *Science and Public Policy*, 22 (4): 218–30.

Tait, J. (1976) *Factors affecting the production and use of pesticides in the UK*. PhD thesis, Cambridge University.

Tait, J., Brown, S. & Carr, S. (1991) Pesticide innovation and public acceptability: The role of regulation. In: *Innovation and Environmental Risk* (eds R. Lewis & A. Weale), Belhaven, London.

Chapter 6

Farmers' Agrochemical Use Decisions and the Potential of Efficiency and Innovation Offsets

Ada Wossink and Tanja de Koeijer

Introduction

The emphasis in this chapter is on the sphere 'agricultural production' as distinguished in the analytical framework outlined in Chapter 1. We discuss the idea that solving pollution problems associated with agrochemical use depends largely on the availability of technical options for more sustainable farming practices. Technical options for farmers in the short and the medium term can be classified in two main categories (see *MacRae et al.* 1990): (1) more efficient use of current agrochemical inputs, and (2) substitution by other, more environmentally friendly inputs. The environmental and financial gains from more efficient use of polluting inputs can be substantial as is highlighted in the empirical agricultural economics literature on technical and environmental (in-)efficiency (see, e.g., Fernandez-Cornejo 1994; Piot-Lepetit *et al.* 1997; Reinhard *et al.* 1999). Regarding 'green' innovation, there is an ongoing debate in the economics literature whether this can potentially combine environmental with economic advantages (Smith & Walsh 2000).

If efficiency improvements and green innovation indeed combine environmental advantages with economic advantages, these offsets would offer a 'free lunch' adjustment to environmental regulations. Policymakers and legislators have at their disposal a number of instruments for controlling agrochemical use such as command and control (e.g. use standards, pollution control equipment requirements, emission limits), economic incentives (fees, marketable permits and liability) and moral persuasion (education). Education in particular is a popular approach to effecting agricultural practices and we discuss whether education is effective in the context of innovation and efficiency offsets.

In the second and third sections of the chapter we provide a theoretical discussion of the economics of the agrochemical use problem. The fourth section presents an empirical analysis of the existence of efficiency and innovation offsets. High-quality survey data for a Dutch sub-sample of sugarbeet producers were used for a case study assessment. The results suggest that there is considerable room for improving environmental quality of agricultural production without conflicting with economic goals. The chapter concludes with a discussion of the policy implications and a summary of the main findings.

Environmental–economic production possibility frontier

Agricultural production generates joint outputs that form two major subsets: food and fibre, and environmental and health effects. The combination in which these marketable outputs and bad side effects are generated is not fixed but depends rather on the production method chosen. Generally, several production methods are available that vary both in their costs and in their environmental impacts. Fig. 6.1 depicts the relationship between agricultural production and environmental impacts for an individual farm in a given natural production environment as defined by climate/weather and soil type and for a given variety of production methods.

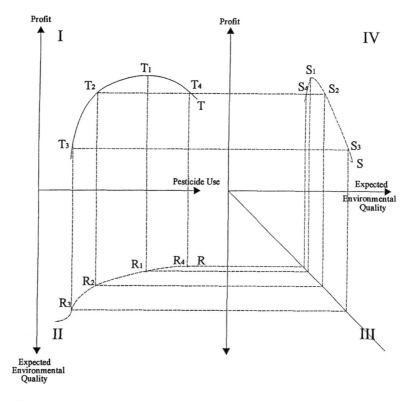

Fig. 6.1 Producer agrochemical use decisions and their effect on expected environmental quality.

The economic relationship between agrochemical use and the producer's profit is illustrated in quadrant I. Every point on the function T shows the maximum amount of profit that can be achieved with a given level of agrochemical use. Alternately, considered from an input orientation, the function describes the minimum amount of agrochemical input required to achieve the given profit level. Without loss of generality, the profit axis could be thought of as the expected utility of profits for risk-adverse producers when there is production uncertainty.

The relationship between agrochemical use and expected environmental quality is represented in quadrant II. Ecosystem health, which is adversely affected by agrochemical use, is represented by function R. The s-shape of this function is derived from the dose–response relationship in toxicology.[1] Quadrant III simply transposes ecosystem health into quadrant IV. Finally, the relationship between ecosystem health and profit is depicted in quadrant IV. This is a production possibility frontier (PPF) that depicts the feasible set of economic performance and ecosystem quality levels.

The shape of the PPF expresses the extent to which economic and environmental performances are compatible. Profits and expected ecosystem health are complements over the increasing range of the frontier and substitutes over the decreasing range. Where markets for environmental services are missing, the larger part of the production possibility frontier is steeply downward sloping as with the PPF in Fig. 6.1 (see Aldy *et al.* 1998). Without any regulation, the economic optimal point is at S_1, with profits T_1 and environmental quality R_1. Obviously, it would be costly to improve the environmental quality of agricultural production to a level beyond R_1.

The presentation in Fig. 6.1 assumes optimal, profit-maximising behaviour of agricultural producers and a given technological state of the art. In practice there will be inefficiency in production and progress in production technology through innovations. The next section analyses the impacts of inefficiency and innovations on environmental improvements and the associated costs.

Efficiency and innovation offsets, education and regulation

While privately owned farms are likely to be efficient with regard to conventional input–output productivity, there are several reasons why there would be inefficiencies in environmental performance: for example, an internal lack of economic incentive and information, bounded rationality, and an absence of external competitive pressure applying to environmental performance. Producers commonly face varying degrees of uncertainty in many aspects of production. For a given production technology, lack of information about the production frontier may lead producers to use inputs inefficiently. Producers may also have limited knowledge of the set of alternative production technologies that are available and their economic and environmental characteristics, as well as a lack of information about how their actions affect environmental quality (Ribaudo & Horan 1999). When in spring farmers decide on the application level of nutrients, they do not know (and cannot accurately predict) the weather for the coming cropping season. As nitrogen fertiliser is cheap and the benefit of applying it is a higher yield and thus a higher return for the crop, farmers will be inclined to supply ample nitrogen in order not to reduce the attainable yield. The same applies to the use of biocides and the occurrence of pests, weeds and diseases. So, farmers make their decisions *ex ante*, whereas evaluation of these decisions is *ex post*.

The importance of inefficiencies for profit and environmental quality is illustrated by farm A_1 in Fig. 6.2. The technology available to producer A_1 is represented by PPF, which is a stylised version of the downward sloping part of S in Fig. 6.1. Profit P_1 and environmental quality W_1 represent the skill with which producer A_1 is currently using the technology. Efficiency offsets available to farm A_1 are along the portion BK of PPF. Points along the lower part of PPF do not provide offsets because profit would decrease. Farms like A_2, which utilises available production technologies efficiently, will likely be on, or close to, the y axis.

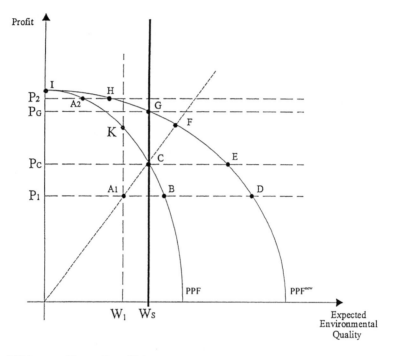

Fig. 6.2   Efficiency and innovation offsets.

Suppose that the socially desired level of expected environmental quality is at W_s. By educating farmer A_1 about the frontier where profits are higher for each level of input use, the producer could be encouraged to use existing management practices more efficiently or to adopt alternative ones. Once on the frontier, the producer could operate according to Good Agricultural Practice (point C) which would achieve the environmental quality goal and at the same time increase profits. However, without any regulation, competition will drive the producer to operate on CI. The expected environmental quality levels that correspond to the production possibilities to the right of K would be an improvement over the initial situation with production at A_1 but do not meet the standard. The environmental quality levels associated with the production possibilities on the portion IK of PPF would even be less than in the original situation. This makes it possible that education about production practices might even reduce environmental quality. Thus, educational assistance and technical innovation alone are not necessarily sufficient to ensure that environmental quality goals are met.

Now assume that a regulation is implemented that specifies the maximum amount of pollution at W_s. Efficiency offsets available to farm A_1, given a standard of W_s, are now along the portion BC of the PPF. For farm A_1 the regulation would entail no compliance cost, since this farm can meet the standard by using its efficiency offsets. Efficiency offsets are not available for (the already efficient) farm A_2 and it will encounter compliance cost of $P_2 - P_C$. However, Hicks' induced innovation hypothesis says producers will seek out technologies that lower the compliance costs of the regulation (Hicks 1963) and the Porter hypothesis asserts that such innovations will be induced by regulation (Porter 1991). The Porter argument is that tough environmental regulation can trigger innovation that may eventually increase a firm's competitiveness and may outweigh the short-run costs of this regulation (Porter & van der Linde 1995; Xepapadeas & de Zeeuw 1999). Assume new technologies become available after some time and expand the production opportunities to a new frontier. Farm A_2 will shift out to this new frontier PPF^{new} and depending on the shape of this new frontier, the innovations will partially offset the environmental compliance costs. If farm A_2 positions itself at H on PPF^{new} it will reduce the cost of complying with the new standard from $P_2 - P_C$ to $P_2 - P_G$. For farm A_1 innovation offsets expand the already existing efficiency offsets from BC to DG.

In summary, the opportunity of the 'free lunch' adjustment offered by efficiency and innovation offsets depends on: (1) the positioning of a farm with respect to PPF^{new}, (2) the shape of PPF^{new}, and (3) the level at which the environmental standard is set.

Empirical assessment of potential efficiency and innovation offsets

Background

Since 1979, research has been carried out in The Netherlands into developing integrated farming systems for field crops. This technical development has brought different possibilities to the farmer to reduce the input of agrichemicals. Most of the studies evaluating the new farming practices concluded that it could benefit farmers through reduced costs of farm inputs, significantly lower emissions of pesticides and nutrients while maintaining yield levels and yield quality (Wijnands 1992). Extension projects were set up to introduce the new methods into farming practice. The pilot farms achieved significant reductions in pesticide and nitrogen use and also achieved attractive financial results. The projects' results had significant consequences, particularly for pesticide policy.

A pesticide use reduction plan was approved in 1991 in The Netherlands (LNV 1991). Reduction targets for pesticide use were formulated as percentage reduction by 1995 and 2000 for the various categories of pesticides compared to the national use in kg ai over the period 1984–1988 (Falconer 1998; Wossink & Feitshans 2000). No standard of use or pollution tax were set to reduce pesticide use. Instead, compliance was arranged largely through a combination of voluntarism, and advice and education. The Dutch farmers' organisation signed an agreement ('convenant') with the government in May 1993 that committed them to achieve the reduction goals, specifically those for crop farming. Applicator training and certification became required for all applicators, and since 1996 application equipment testing is required for all equipment.

Based on the theoretical model described above, two hypotheses were derived for the situation outlined. First, we expect to find offsets offered by improvements in efficiency and by the introduction of less environmentally harmful pesticides. Second, we expect the environmental improvements offered by the offsets to be rather unutilised because of a lack of regulation.

Data

To test for the existence of efficiency and innovation offsets, high quality survey data[2] were used describing sugarbeet production of 111, 116, 119 and 121 farmers situated in Flevoland for the years 1994, 1995, 1996 and 1997 respectively. The region was selected for its distinct natural geographic boundaries and data availability. The farms are located in a 50 by 50 km region and it was therefore assumed that all farmers experienced identical physical conditions and therefore produce under the same production possibility frontier in a given year. Data were available for almost all farmers for each year of the period 1994–1997 and so the performance of the farmers over different years could be compared.

For the measurement of the offsets, two variable inputs were used, nitrogen fertiliser and herbicides. These two inputs can be regarded as the two most important inputs with respect to environmental pollution caused by the cultivation of sugarbeet. The nitrogen input was measured as the sum of the amount of nitrogen fertiliser, the amount of nitrogen available in the soil in spring[3] and the amount of nitrogen which becomes available during the cropping season by mineralisation of organic nitrogen expressed in kg N per hectare. The environmental impact of nitrogen was expressed in the amount of nitrogen surplus (kg N) per ha. The latter is measured as the difference between the amount of total nitrogen input and the amount of nitrogen in the product. The environmental effect of herbicides was expressed in environmental impact points (EIP). These EIPs are based on three environmental effects: leaching into groundwater, effects on water organisms and effects on soil organisms. For the measurement of these EIPs an environmental yardstick has been developed (Reus & Pak 1993). The environmental yardstick version 1997 was used in the DEA estimations (Kerngroep MJP-G 1997).

For the economic performance, gross margin[4] per crop was assessed instead of profit per farm because only one crop was analysed. In doing so it was assumed implicitly that: (1) the optimal amount of 'farm-specific' or 'fixed' inputs like labour and machinery required for sugarbeet production is not conflicting with the optimal use of these 'fixed' inputs considering the total cropping plan of the farm, and that (2) improvements in technical farm performance do not affect fixed costs. The variation in gross margins of sugarbeet is, apart from yield variation, mostly due to differences in the costs for nitrogen and herbicides and only these two inputs were taken into account. Improvements in nitrogen use can be achieved by better attuning input level to expected yield levels. For herbicides use opportunities are offered by 'low dosage' systems (spraying a low quantity of chemicals just after the emergence of the weeds) or by selecting herbicides that are less environmentally harmful, or both. The monetary yield was calculated with the price per net ton of sugarbeet corrected for the sugar content, the extractability index, tare and a premium for early or late delivery to the sugar processing industry.

Empirical method

Data Envelopment Analysis (DEA) was employed to quantify the offsets and the extent to which an individual farmer employs these. DEA constructs a frontier representing the latest technology and simultaneously calculates the distance to that frontier for the individual observations (see, e.g., Paris 1991; Färe *et al.* 1994; Tyteca 1996). The frontier is piecewise linear and is formed by tightly enveloping the data points of the observed 'best practice' activities in the observations, that is, the most efficient and innovative farms in the sample. So it is assumed that the performance of the best farmers can be used to assess a benchmark for the state of the art PPF^{new}.

DEA uses the distance to the frontier as a measure of efficiency[5]. In Fig. 6.2, farm A_1 is compared to point F on the frontier PPF^{new} to calculate the total of innovation and efficiency offsets available to this specific farm. The comparison results in an offset use efficiency of OA_1/OF. Differences in the distance to the frontier provide a score for each farm from 0 (worst performance) to 1 (best performance). A low score indicates considerable unused offsets for the specific farm whereas a score of 1 indicates that the farm is located on the frontier[6].

In addition to the calculation of the offset efficiency, the DEA method was employed to calculate several other efficiency measures for each farm in the sample. These other measures allow the sources of offset inefficiency to be identified. Traditionally, efficiency analysis has focused on marketable output relative to paid inputs (technical efficiency of input use). In the case of two input variables, x_1 and x_2, assessing the input per unit of output provides a plot where the co-ordinates (x_1/y and x_2/y) indicate the technical efficiency of the used inputs. Notice that in this case the points closest to the origin generate a frontier and the set of observed combinations of inputs-per-unit-of-output are on or above this curve. The deviation from the efficiency frontier is considered to be associated with technical inefficiency of the farms involved.

The DEA model for each specific production unit is formulated as a fractional programming problem[7]. For example, the (dual) formulation for the *technical efficiency* of farm *j* is:

Minimise	Φ_j	(1a)
subject to	$Yv_j \geq y_j$	(1b)
	$Bv_j \leq b_j \Phi_j$	(1c)
	$v_j \geq 0$	(1d)

where Φ_j is the measure of technical efficiency of the *j*-th farm; Y is a $p \times n$ matrix of *p* outputs produced by the *n* farms; v_j is the intensity vector of the weights attached to the *n* farms for the construction of the virtual comparison unit for farm *j*; y_j is a $p \times 1$ vector of quantities of output produced by farm *j*; *B* is a $m \times n$ matrix of *m* inputs used by the *n* farms, and b_j is the vector of these inputs for farm *j*. The efficiency of the *n* farms is assessed by solving *n* LP models, in which the vectors y_j and b_j are adapted each time for the farm *j* considered.

The environmental efficiency can be measured in the same way as the technical efficiency. Instead of the amount of observed inputs, the observed environmental impacts are used (Tyteca 1996, 1997). To express the fact that the environmental damage per unit input depends on the area over which the input is spread, the environmental impact is not measured per unit product but per unit area. The model to calculate the (area oriented) *environmental efficiency* of farm *j* is:

Minimise	Δ_j	(2a)
subject to	$uv_j \geq 1$	(2b)
	$Zv_j \leq z_j \Delta_j$	(2c)
	$v_j \geq 0$	(2d)

where Δ_j is the environmental efficiency, *u* is the vector of acreage used for production on the *n* farms; *Z* is a $r \times n$ matrix of *r* environmental impacts generated by the *n* farms and v_j is defined as before. Compared with the model for the agronomic and technical efficiency, restriction (1b) is replaced by the acreage constraint (2b), which ensures that the pollution per unit of area is minimised while searching for the efficient farms.

The economic efficiency can be measured without DEA by comparing realised gross margins per hectare among farms. Let *W* be the gross margin per farm defined as the value of production minus total variable costs. The *economic efficiency, EcE_j*, of the production for farm *j* can then be calculated as:

$$EcE_j = \frac{W_j}{Max_n W_n} \qquad (3)$$

Using the results of the models above, the *offset use efficiency* Ω_j of farm *j* now can be assessed as follows:

Minimise	Ω_J	(4a)
subject to	$Sv_j \geq s_j$	(4b)
	$uv_j \leq \Omega_j$	(4c)
	$v_j \geq 0$	(4d)

where *S* is the matrix of the calculated 'outputs' (environmental and economic efficiency) for *n* farms and *s* is the vector of these calculated outputs for farm *j*.

For a more detailed discussion of alternative DEA models to measure environmental performance, see Tyteca (1996; 1997) and Callens & Tyteca (1999).

Results

Table 6.1 presents the sample mean, the standard deviation and the minimum value of the technical efficiency; the environmental area oriented efficiency, the economic efficiency and the offset use efficiency per field. On average the estimated environmental efficiency is very low (0.40) and the economic efficiency relatively high (0.79). Besides, the standard deviation is considerably lower for the economic efficiency than it is for the environmental efficiency. The estimated environmental efficiency indicates that the environmental impact per unit area can be reduced by 60% by reducing N-fertiliser use and reducing herbicide use or by substituting herbicides with a lower EIP score. The average offset use efficiency was found to be relatively high (0.80). To interpret this result we refer to Fig. 6.2 and recall that the offset use efficiency is measured as the distance to the frontier PPF^{new}. This distance will be small (which means a high offset use efficiency) when the data show only limited variation in one of the two indicators which in this case holds for the economic efficiency.

Table 6.1   The technical efficiency (TE), environmental efficiency (EE), economic efficiency (EcE) and the offset use efficiency (OUE) of sugarbeet growers in Flevoland 1994–1997.

		TE	EE	EcE	OUE
1994[1]	Average	0.50	0.34	0.83	0.84
	Standard deviation	0.17	0.17	0.08	0.08
	Minimum	0.28	0.07	0.55	0.57
1995[2]	Average	0.46	0.33	0.79	0.81
	Standard deviation	0.14	0.17	0.08	0.09
	Minimum	0.22	0.01	0.49	0.49
1996[3]	Average	0.54	0.48	0.79	0.80
	Standard deviation	0.16	0.21	0.09	0.10
	Minimum	0.24	0.07	0.43	0.44
1997[4]	Average	0.47	0.43	0.75	0.76
	Standard deviation	0.15	0.24	0.14	0.13
	Minimum	0.14	0.01	0.00	0.32
Average 1994–1997		0.49	0.40	0.79	0.80

[1]n = 143; [2]n = 138; [3]n = 142; [4]n = 131
Based on data of sugarbeet growers of the Dutch sugar beet processing company Suiker Unie.

The information provided by the DEA analysis can be used to identify management strategies to combine profit objectives with environmental quality. To this end Table 6.2 presents the rank correlation between the efficiency measures. There is a significant correlation for each of the four years analysed between the technical efficiency and the offset use efficiency, economic efficiency and environmental efficiency. These results suggest that farmers who focus on optimising technical efficiency follow a good strategy to achieve economic and environmental efficiency.

Next, farmers' performance was analysed over the years in the data set to test whether differences in management can explain inefficiency. The Spearman ranking showed significant rank correlation for the various efficiencies measures over the

Table 6.2   Spearman rank correlation between efficiency measures for 1994, 1995, 1996 and 1997 (based on data of sugarbeet growers of the Dutch sugarbeet processing company Suiker Unie).

Rank correlation		1994	1995	1996	1997
OUE	TE	0.42**	0.49**	0.55**	0.67**
EcE	TE	0.30**	0.37**	0.42**	0.60**
EE	TE	0.77**	0.65**	0.61**	0.48**

* Significant at P < 0.05 level; ** Significant at P < 0.01 level.

1994–1997 period except for the environmental efficiency and the economic efficiency of 1994 and 1997, and for the economic efficiency and the offset use efficiency of 1994 and 1995 (Table 6.3). As we assume that all farmers operated under the same physical conditions, this indicates that there are persistent differences in management quality among the farmers in our sample. The correlations suggest that substantial improvement in environmental performance could occur with accompanying economic benefits.

Table 6.3   The average rank correlation (Spearman) for the efficiency measures 1994–1997 for the same farmer (based on input and output data of sugarbeet growers of the Dutch sugarbeet processing company Suiker Unie).

Correlation of		1997[1]	1996[2]	1995[3]
TE	1994	0.51**	0.45**	0.60**
	1995	0.57**	0.53**	
	1996	0.43**		
EE	1994	0.17	0.33**	0.65**
	1995	0.22*	0.34**	
	1996	0.39**		
EcE	1994	0.19	0.23*	0.16
	1995	0.44**	0.36**	
	1996	0.26**		
OUE	1994	0.21*	0.23*	0.17
	1995	0.43**	0.39**	
	1996	0.26**		

[1] n = 97, 101, 106 respectively; [2] n = 106, 111 respectively; [3] n = 104.
*Significant at P < 0.05 level; ** Significant at P < 0.01 level.

Conclusions and policy implications

The objective of this paper was to discuss and analyse empirically the implications of efficiency and innovation offsets for the management of non-point source pollution from agriculture. Based on the theoretical economic model, two hypotheses were derived and tested for the case of pesticide and nitrogen use in sugarbeet production. First, we expected to find efficiency and innovation offsets. Second, we expected the environmental improvements offered by the offsets to be relatively unutilised compared to the economic improvements because of a lack of environmental regulation.

Substantial heterogeneity was found in performance. The average offset use efficiency was 0.80; however, the environmental efficiency of the farmers in our sample was only 0.40 and the average technical efficiency was only 0.49 with significant and persistent differences among farmers over the years. A significant positive rank correlation was found between the technical efficiency and the offset use efficiency and also between the technical efficiency and the economic and environmental efficiency. These results suggest that there is considerable room for controlling the non-point source pollution problem of agricultural production without conflicts between economic and environmental goals; the key factor is the improvement of technical efficiency.

The differences in efficiency among farmers were found to persist within years (over fields) and also between years. As physical conditions could be assumed to be fairly similar for all farmers in the data set, differences in efficiency must be mainly the result of differences in farm management. In order to know how efficiency can be improved in practice, it is essential to know the factors determining managerial success. Few studies have analysed the relationship between the total complex of farm management factors and technical farm performance. De Koeijer *et al.* (2001) investigated this relationship for nitrogen management on Dutch crop farms and identified a lack of knowledge about how practices affect environmental quality as an important reason for environmental inefficiency. This result is in line with those of Baarda (1998) who concluded that farmers found it difficult to estimate the economic and environmental effects of changing practices but that extra information on these relationships was very useful to improve decision making. However, there is no guarantee that education alone will be effective in achieving an improvement in environmental quality and this is supported by the case study results. In general, the outcome of educational programmes depends on the profit–environmental quality frontier compared with farmers' initial understanding of this frontier. The average environmental efficiency (only 40%) shows that environmental improvements offered by innovation and efficiency offsets are only very partly utilised. In contrast, the average economic efficiency was 79%.

The case study results indicate that in a situation of innovation and inefficiency, education's value may be limited to being a component of a policy mix that includes economic incentives (taxes on pesticides) or direct regulation to ensure that the agricultural sector will provide the environmental services desired by society. From the perspective of the problem setting outlined above, these outcomes demonstrate the complexity of factors affecting farmers' agrochemical use decisions. More specifically they demonstrate the interaction of actual, i.e. non-optimal, producer behaviour, agricultural innovation and the design of effective socio-economic regulations.

Acknowledgements

The authors thank Frank M. Wefering for research assistance with this paper. The use of data provided by the sugarbeet processing company Suiker Unie is gratefully acknowledged.

Notes

1 An important assumption in this relationship is that there is always a dose below which no response occurs or can be measured. A second assumption is that once a maximum response is reached any further increase in the dose will not result in any increased effect.
2 Unitip data provided by the sugarbeet processing company Suiker Unie.
3 About 25% of the observations missed data on the amount of nitrogen in the soil in spring. In those cases it was assumed that this amount was equal to the average amount of nitrogen available in that particular year. In 1994, 1995, 1996 and 1997 the average amount of nitrogen in the soil was 27, 27, 88 and 49 kg N per hectare respectively.
4 Defined as returns minus variable costs.
5 The distance to the frontier is measured in various ways depending on assumptions regarding the production technology and the objective of the study (Färe *et al.* 1994). When production technology is homothetic, expansion paths are rays from the origin and recommendations for radial adjustments are consistent with achieving efficiency. In the analysis presented here the focus is on variable inputs only and a radial measure was used.
6 Notice that if there is little variation in the performance of the units they all will have relatively high scores.
7 A clear overview is given in Paris (1991).

References

Aldy, J.E., Hrubovcak, J. & Vasavada, U. (1998) The role of technology in sustaining agriculture and the environment. *Ecological Economics*, 26, 81–96.

Baarda, C. (1999) *Politieke besluiten en boerenbeslissingen: Het draagvlak van het mestbeleid tot 2000*. PhD thesis, Rijksuniversiteit Groningen.

Callens, I. & Tyteca, D. (1999) Towards indicators of sustainable development for firms: A productive efficiency perspective. *Ecological Economics*, 28, 41–53.

Falconer, K.E. (1998) Managing diffuse environmental contamination from agricultural pesticides: An economic perspective on issues and policy options, with particular reference to Europe. *Agriculture, Ecosystems and Environment*, 69, 37–54.

Färe, R., Grosskopf, S. & Lovell, C.A.K. (1994) *Production frontiers*. Cambridge University Press, Cambridge, MA.

Fernandez-Cornejo, J. (1994) Nonradial technical efficiency and chemical input use in agriculture. *Agricultural and Resource Economics Review*, 23, 11–21.

Hicks, J.R. (1963) *The Theory of Wages*. St Martins Press, New York.

Kerngroep MJP-G (1997) *Milieumeetlat rekenprogramma*. Kerngroep MJP-G, Ede.

de Koeijer, T.J., Wossink, G.A.A., Smit, A.B., Janssens, S.R.M., Renkema, .J.A. & Struik, P.C. (2001) A methodology to assess quality of farmers' environmental management and its effects on resource use efficiency illustrated for Dutch arable farmers (submitted).

LNV (1991) *Meerjarenplan Gewasbescherming.* Tweede Kamer, vergaderjaar 1990–1991, 21677 nos. 3–4. Staatsuitgeverij, The Hague.

MacRae, R.J., Hill, S.J., Henning, J. & Bentley, A.J. (1990) Policies, programs, and regulations to support the transition to sustainable agriculture in Canada. *American Journal of Alternative Agriculture*, 5, 76–89.

Paris, Q. (1991) *An Economic Interpretation of Linear Programming*. Iowa State University Press, Ames.

Piot-Lepetit, I., Vermersch, D. & Weaver, R.D. (1997) Agriculture's environmental externalities: DEA evidence for French agriculture. *Applied Economics*, 29 (3), 331–42.

Porter, M.E. (1991) America's green strategy. *Scientific American*, 264 (4), 168.

Porter, M.E. & van der Linde, C. (1995) Towards a new conception of environment–competitiveness relationships. *Journal of Economics Perspectives*, 9, 119–32.

Reinhard, A.J., Lovell, C.A.K. & Thijssen, G.J. (1999) Econometric estimation of technical and environmental efficiency: an application to Dutch dairy farms. *American Journal of Agricultural Economics*, 81, 44–60.

Reus, J.A.W.A. & Pak, G.A. (1993) An environmental yardstick for pesticides. *Mededelingen van de Faculteit Landbouwwetenschappen, Rijksuniversiteit Gent*, 58 (2a), 249–55.

Ribaudo, M.O. & Horan, R.D. (1999) The role of education in non-point source pollution control policy. *Review of Agricultural Economics*, 21 (20), 331–43.

Smith, V.K. & Walsh, R. (2000) Do painless environmental policies exist? *Journal of Risk and Uncertainty*, 21 (1), 73–94.

Tyteca, D. (1996) On the measurement of the environmental performance of firms: A literature review and a productive efficiency perspective. *Journal of Environmental Management*, 46, 281–308.

Tyteca, D. (1997) Linear programming models for the measurement of environmental performance of firms. *Journal of Productivity Analysis*, 8 (2), 183–98.

Wijnands, F.G. (1992) Evaluation and introduction of integrated arable farming in practice. *Netherlands Journal of Agricultural Science*, 40, 239–49.

Wossink, G.A.A. & Feitshans, T.A. (2000) Pesticide policies in the European Union. *Drake Journal of Agricultural Law*, 5 (1), 223–49.

Xepapadeas, A. & de Zeeuw, A. (1999) Environmental policy and competitiveness: The Porter hypothesis and the composition of capital. *Journal of Environmental Economics and Management*, 37, 165–82.

Chapter 7
Variations in Agricultural Practice and Environmental Care

Geert R. de Snoo

Introduction

The main traditional focus of environmental policy on pesticides has been a progressive tightening of the criteria employed for compound approval. With the establishment of criteria for non-target species toxicity, biodegradability, mobility, lipophility and so on, a great many first-generation crop protection chemicals have now disappeared from Western, industrialised markets. A second policy concern has been to reduce aggregate pesticide use and in many countries volume consumption has indeed fallen substantially. In The Netherlands, a country with one of the world's highest levels of per-hectare pesticide use, national and sectoral reduction targets were established (MJP-G 1991). By the year 2000, aggregate Dutch pesticide use had been reduced by 50% relative to the 1990 policy baseline (the average annual use over the period 1984–1988), mainly through a mandatory extension of crop rotations in tandem with strict controls on the use of soil fumigants (de Jong *et al.* 2001).

It has become increasingly clear, however, that this twofold strategy of volume policy and tighter registration no longer suffices for continued improvement of rural environmental quality, but that the intricacies of field pesticide application must also be addressed. Examples of recent regulation include the growing requirements being set on farmers' knowledge of pesticides in the context of permits, periodical approval of spraying equipment, use of low-drift spray nozzles and implementation of buffer zones along watercourses and certain other areas. In the process of focusing on actual pesticide application this chapter is encroaching ever further on the everyday operations of the individual farmer. While national generic consumption data and approval criteria proved sufficient for pursuing national policy targets, continued improvement of environmental quality requires enhanced knowledge and understanding of the activities of individual farmers.

In this chapter we therefore look more closely at field pesticide use, the environmental impact of that use and the scope for effective environmental protection. In doing so, our main concern will be the empirical variation observed in agricultural practice, the reasoning being that, if there is wide variation in use and impact among comparable agricultural holdings, variation may provide leverage for reducing pesticide use and enhancing environmental quality.

Observed variation in agricultural practice

To gauge pesticide use on arable and horticultural holdings a case study was undertaken in The Netherlands, employing for this purpose a survey of 3200 Dutch farmers carried out in 1995 by Statistics Netherlands (CBS 1997). In this survey quantitative information was collected on crop chemical use on the main national arable and horticultural crops. The survey did not cover the use of soil fumigants.

The survey points to considerable variation in pesticide use among farmers. As an illustration, Table 7.1 shows average pesticide use on The Netherlands' major arable crops and the deviation of this overall average use of the top and bottom 20% brackets of the observed range from this average. The range covered by these two indices has here been taken as a measure of variation in national use, with higher top-end values indicating greater variation from the national average. In this scale average variation in pesticide use across the various agricultural crops (including arable, fruit, bulbs, etc.) was 4–2.3 in 1995 (de Snoo & de Jong 1999). As an approximation, then, in The Netherlands the intensity of pesticide use on farm holdings in the bottom 20% bracket of users is about one-quarter of the national average, while that of the top 20% is over twice that average. Earlier surveys of arable farmers and nursery and greenhouse operators show that this variation has remained fairly constant over the years (CBS 1994).

Table 7.1 Average pesticide use on principal Dutch arable crops in 1995, in kg active ingredient ha^{-1}, showing deviation of bottom and top 20% user brackets from national average (CBS 1997).

Sector/Arable crop	Average use (kg ai/ha)	Deviation of bottom 20%	Deviation of top 20%
Ware potatoes	11.6	3.3	1.8
Industrial potatoes	10.6	2.9	1.6
Seed potatoes	20.8	3.7	3.4
Sugarbeet	3.6	2.6	2.0
Fodder maize	3.2	3.2	1.9
Winter wheat	2.7	3.4	1.8

A comparison of individual agricultural and horticultural crops shows that variation is particularly high for tree and nursery stock (10–4.9), although it should be borne in mind that these are multicrop systems. By way of illustration, Fig. 7.1 shows inter-holding variation for four different crops. As can be seen, some tulip growers use five to ten kilogram active ingredient per hectare (ai ha^{-1}), while others apparently require over fifty kg ai ha^{-1} in the same season. The pattern of variation varies widely from crop to crop.

This degree of farm-to-farm variation in pesticide use is not peculiar to The Netherlands but is common throughout Europe. Fig. 7.2 shows pesticide use (in kg ai ha^{-1}) in four European regions including the Dutch Flevopolder, based on a 1994 survey of sixty potato farmers in each region. The high level of pesticide use in north-west France is striking, a result of particularly heavy use of fungicides. In contrast to the

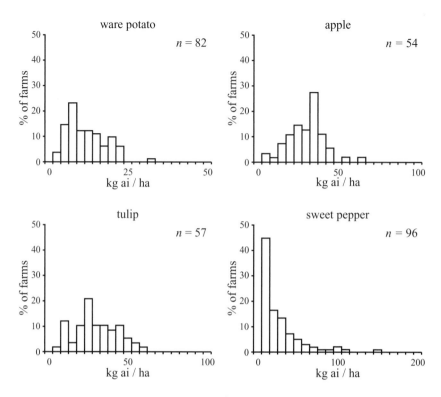

Fig. 7.1 Variation in farm pesticide use (kg ai ha^{-1}) on four crops in 1995 (CBS survey data 1995). N = no of farms.

Flevopolder, for example, which has a broadly similar rotation scheme, the same share of ware potatoes and the same potato varieties, in northwest France mancozeb is the preferred agent for controlling potato blight, as against fluoazinam, which requires a lower volume dosage (EC 1996). Farm-to-farm variation in pesticide use is also high for a number of other crops, including winter wheat, wines and apples (EC 1997).

Causes of inter-farm variation

Inter-farm variation in pesticide use may be caused by a wide variety of factors. In the first place there are regional differences in terms of climate, soil and the incidence of weeds, pests and diseases, in relation to the selected crop varieties, etc. At the same time, though, major variation is also observed among holdings within one and the same region, where uniform conditions of soil and climate prevail (de Snoo *et al.* 1997). Ultimately, quantity of pesticide product applied by individual farmers is determined by three basic factors:

- choice of product: some products are far more biologically active and require a lower kilogram dosage;
- frequency of application: some farmers are quicker to spray than others, depending on their risk perception (costs/benefits, etc.);

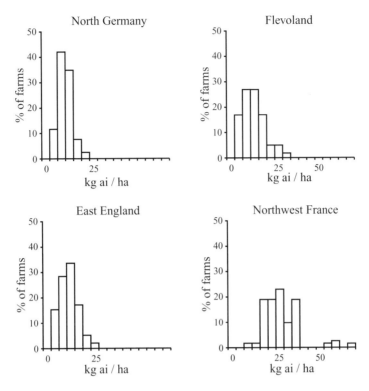

Fig. 7.2 Variation in pesticide use (kg ai ha⁻¹) on ware potato crops in four European regions in 1994 (EC 1996). In each region 60 farms were investigated.

- quantity per application, i.e. the field dose employed.

The impact of each of these three factors varies, depending on the crop and in particular on the range of crop protection agents available on the market and the preferred control strategy (Buurma 1997). The variation in field dose is of specific concern. Recommended pesticide dose rates are indicated both on packaging labels and in numerous farm handbooks and these are not meant to be exceeded; indeed, farmers may use less at their discretion. High dose rates may lead to excessive concentrations in harvestable crops: an undesirable state of affairs from a public health perspective. They may also pose significant environmental risks. It is commonly held that 'educated Western farmers' use pesticides sparingly, often below the recommended dose rates. With the empirical data of Statistics Netherlands for 1995 at hand (CBS 1997), the truth of this statement can be assessed by comparing observed variation in use with recommended doses. By way of example, Figure 7.3 shows dimethoate use on ware potato crops for aphid control. As can be seen, *average* use of this chemical is already above the recommended level (0.25 *versus* 0.20 kg ai ha⁻¹). Fig. 7.3 also illustrates that while the average overdose may be relatively limited (50 g ai ha⁻¹⁾, in practice this means that a very large proportion of farmers are exceeding recommended dose rates. This is not counter-balanced by those farmers employing doses below recommended rates.

Table 7.2 aggregates the data across crops, reviewing the situation for the most

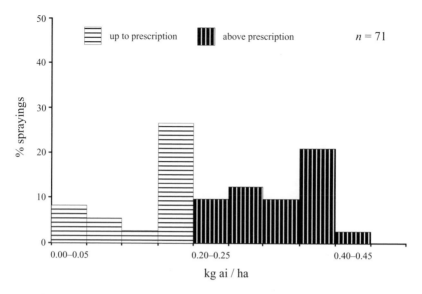

Fig. 7.3   Variation in dimethoate dose (kg ai ha⁻¹) in ware potato crops in 1995, among 71 sprayings (*N*) to control aphids (CBS survey data 1995).

frequent pesticide applications, i.e. products used to control a specific weed, pest or disease in a specific crop. Table 7.2 shows that, on average across crops, fungicides and insecticides are used above the recommended dose in one in every four or five custom applications. In the case of herbicides this figure is considerably lower, namely about once every ten applications. *Average* field doses of fungicides and insecticides (i.e. including holdings using below-recommended doses) are respectively 2% and 6% higher than recommended. For herbicides the average dose is 16% lower than recommended. The relatively wide spread of values around this average is due to one specific product used in two crops; if this application is ignored, average herbicide use is 33% below the recommended dose.

Table 7.2   Average percentage of high-frequency pesticide applications (25 times used in 1995, at least, to control a specific weed, pest or disease in a specific crop), implemented at above recommended dose, and percentage deviation from this dose, with standard deviations (CBS survey data 1995).

	Number of applications	Applications above recommended dose (%)	Average deviation from recommended dose (%)
Fungicides	101	21.6 ± 17.2	1.8 ± 50.6
Insecticides	86	22.6 ± 17.4	6.2 ± 50.1
Herbicides	94	10.6 ± 13.9	−15.8 ± 123.6

The relative infrequency by which recommended herbicide dose rates are exceeded may be due to the fact that over-dosage of weed control chemicals may lead to crop damage, which is not the case with fungicides or insecticides. The risks associated with herbicide under-dosage are relatively minor, on the other hand, for a certain amount of corrective action can be taken in the form of a repeat spraying or mechanical weed control. In addition, herbicides may be used on only some part or parts of a given plot, down individual crop rows, for example. In the case of insecticides (including these for insect-borne viruses) and fungicides, spraying is generally undertaken as a preventative action and under-dosage carries the risk of major harvest losses.

Given the fact that crop chemicals have been tested for efficacy at the recommended dose rate in the approval stage, there should be no need for over-dosage. The interesting question is therefore: why does over-dosage occur? A closer scrutiny of the 1995 data, and in particular those applications in which the highest overdoses occur shows that two chief forms of over-dosage can be distinguished.

In the first place there is 'erroneous use'. This form of over-dosage may result from errors of calculation (using precisely ten times the recommended dose, for example) or from an inadequate grasp of appropriate usage (when different doses hold for indoor and field application, for example). Relatively new products that are to be applied in minute doses are also particularly prone to over-application. Here, psychological factors may also be at work: it may be hard for the farmer to believe that where kilogram doses were once the rule, a tenth of a gram will now suffice. This form of over-dosage might be restricted by marketing less concentrated product formulations.

Over and against 'erroneous' over-dosage, there also appears to be 'deliberate' application of pesticides at above recommended doses. In many cases a large proportion of farmers are guilty of such malpractice. As pointed out above, this is particularly true of fungicides and insecticides. In the latter case spraying costs also appear to be a factor. Insecticide use is relatively inexpensive (generally 50 Euro or less per hectare) and so, therefore, is over-dosage. In the case of expensive insecticides, application is invariably below the recommended dose. A similar pattern emerges for fungicides, but it is less pronounced. Herbicides, as already mentioned, are often relatively expensive in terms of per-hectare requirements, and breaches of recommended dose rates are observed only among the products that are cheaper (less than about 50 Euro per hectare).

This variation in pesticide use certainly merits further study. A more detailed analysis might then be made of how this phenomenon correlates with the area to be sprayed, for example, and with cost-benefit considerations at both crop and farm level. This might answer the related questions of whether introduction of a levy on relatively inexpensive plant protection products would encourage users to adhere more strictly to recommended dose rates and whether this might then serve as an effective policy instrument for reducing pesticide use (see Oskam *et al.* 1998).

Although volume pesticide use is not in itself a one-to-one indicator of potential environmental damage, the observed variation in farm-level consumption does offer leverage for environmental protection policy. Although current pesticide approval procedures make due allowance for variation in ecological conditions and the sensitivity of non-target organisms, this is not the case for 'variation in farmer behaviour', a parameter that might usefully be included in risk assessment. Pesticides are, after all, one of the few classes of hazardous substances that are deliberately introduced into the environment and the human element is, thus, always very much to the fore. Instead of basing risk estimates on the assumption that farmers will use no more than the maximum recommended dosage ('good agricultural practice'), it might be preferable to assume 'standard agricultural practice', i.e. to include the risk of both deliberate and erroneous over-dosage. As an initial step, calculations might be based on a spread in farmer behaviour representing 95% of farmers, an approach in line with current thinking on probabilistic risk analysis (see Solomon 1999).

Environmental impact

Having considered the empirical variation in kilogram use of pesticides, the question is now whether any correlation can be established with environmental impact. The kilogram quantity of product required for effective control may after all vary substantially. In addition, the field impact of a pesticide on non-target organisms also depends on the mode of application, the properties of the active ingredient and a wide range of environmental factors. At present a great many different methodologies are employed in Europe for estimating the environmental impact of pesticide use (Reus *et al.* 2002). To arrive at a best estimate of overall pesticide impact in The Netherlands we here employ the 'environmental yardstick' developed in this country. This yardstick is designed to assess the potential impact of individual, plot-level applications and has been elaborated for virtually all the active ingredients and pesticide formulations currently on the market (IKC 1996; CLM 1997). The yardstick assigns 'environmental impact points' (EIP) to a given application of a pesticide formulation, reflecting its potential environmental impact: the higher the score, the greater the risk. EIP are assigned for three categories of environmental impact: risks to aquatic and soil biota and risk of groundwater pollution due to leaching. The yardstick has been designed so that an application scoring 100 EIP or less for one category satisfies official Dutch and European pesticide approval criteria.

To assess the overall environmental impact of Dutch pesticide use, then, the environmental yardstick was applied to the empirical farmer data set, using standard Dutch surface water emission factors and regional soil characteristics to calculate the emissions associated with each individual pesticide application (see de Snoo & de Jong 1999).

Kilogram use and environmental impact

As noted, there is no universal relationship between kilogram pesticide use and environmental impact. There is an enormous range in required application rates, and the same holds for toxicity, method of application and environmental fate. By way of illustration, Fig. 7.4 shows how average per-hectare insecticide use (in kilograms ai) in winter wheat correlates with average environmental impact points. As the example shows, lower volume usage may not necessarily lead to less environmental impact if substitute products are more potent. Different results are found for the other categories of pesticide and for other crops. Alternative methods of estimation confirm that there is no universal relationship between volume pesticide use and potential environmental impact (Reus *et al.* 2002).

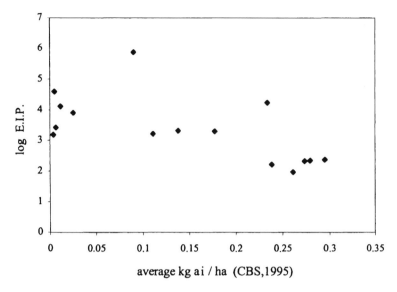

Fig. 7.4   Potential environmental impact (log) as a function of kilogram insecticide use in winter wheat in 1995. Each dot indicates average use of a specific compound per treatment.

A comparison of the respective shares of insecticides, herbicides and fungicides likewise demonstrates the absence of a correlation between volume use and environmental impact. Insecticides, though representing only about 5% of Dutch aggregate pesticide use (in kg), account for some 40% of overall environmental burden. Herbicide use, while substantially higher, has considerably less environmental impact. Fungicides, the group of compounds most intensively used, contribute about 42% to the overall burden (Fig. 7.5). A crop-by-crop comparison shows that in most crops insecticide use accounts for the bulk of environmental impact. The major exceptions are ware and industrial potatoes, where fungicides are clearly the key contributor.

Turning to the three environmental compartments–soil, water and air–a cross-crop analysis for The Netherlands shows that by far the greatest impact associated with pesticide use is on aquatic ecosystems: 91% of overall impact as gauged with the

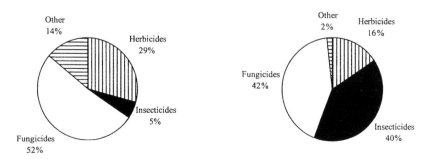

Fig. 7.5   Use of pesticide categories in The Netherlands in 1995 (a), and share of environmental impact based on EIP (b).

environmental yardstick. The remainder accrues almost evenly to the other two categories of impact measured by the yardstick: impact on soil biota (5%) and groundwater pollution (4%).

Fig. 7.6 shows the environmental burden *per hectare* of the main agricultural sectors. As can be seen, there are major sectoral differences. Bulb-growing has the greatest environmental impact. By multiplying the per-hectare environmental burden associated with each crop by national crop area, the *aggregate* environmental impact of the various sectors can be compared (Fig. 7.6). In The Netherlands, on this basis arable farming has by far the greatest overall environmental impact, with the various potato crops on their own accounting for just over half the aggregate environment impact of pesticide use in this country.

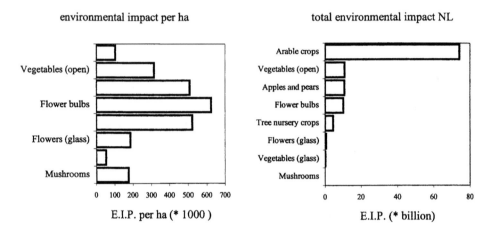

Fig. 7.6   Environmental impact in various sectors in 1995: per hectare (a), for The Netherlands as a whole (b).

Intra-crop variation in environmental impact

As an example, Fig. 7.7 shows the variation in environmental impact occurring within several specific crops. There is a substantial spread and it does not always follow the same statistical pattern. In the case of apples and sweet peppers, environmental impact has a roughly *normal* distribution, but this does not hold for the other two crops. In the case of ware potatoes there are a relatively large number of growers scoring between 100 000 and 300 000 points on the EIP scale but also a considerable proportion causing even greater environmental impact. This pattern is still more pronounced for tulip cultivation, which shows the greatest spread in environmental impact.

Causes of variation in environmental impact

Inter-holding variation in environmental impact is due in part to the variability of pesticide use associated with such agricultural factors as differences in pest and disease incidence. However, the results of a pilot study demonstrate that there may be

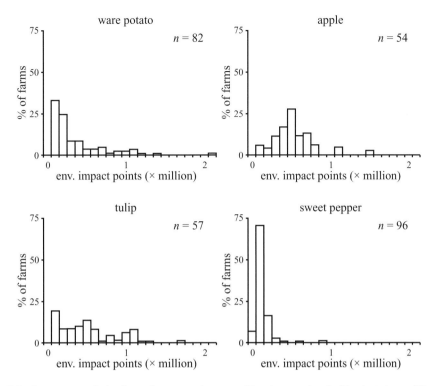

Fig. 7.7   Intra-crop variation in per-hectare environmental burden associated with selected crop (N = no. of farms).

substantial variations in impact within one and the same crop grown within a limited geographic region (de Snoo *et al.* 1997). In all likelihood, then, differences in choice of product and method of application (aerial or field spraying, etc.) are a far more important contributing factor. The cited study indicates that environmental impact is not correlated with such factors as holding size or farmer age and education. Neither could any clear correlation with crop yields be observed; the holdings with the highest yields do not automatically have the greatest environmental impact. The potential environmental impact of products was found to have virtually no influence on farmers' choices of product (de Snoo *et al.* 1997).

Environmental prevention

Practical experience gained in a variety of major crops in The Netherlands demonstrates that very substantial reductions in the environmental impact of farm operations can be achieved. This is borne out by a number of demonstration projects. The environmental impact of potatoes grown under a 'green label' (*Agro-milieukeur*) regime is a factor of 50 less than that of conventionally grown potatoes (de Jong & de Snoo 2002). This environmental gain is due to a combination of product choice, buffer zones and technical measures such as the use of low-drift spray nozzles. In other crops, too, a suitable combination of cropping measures, choice of product and other enlightened action can markedly reduce not only pesticide emissions but also the environmental damage to which these give rise.

With the environmental impact data for individual Dutch farm holdings, calculations can be run to estimate the effectiveness of specific abatement measures. This allows, for example, the effect of tackling the most polluting holdings to be assessed. This can be done by setting the environmental impact of the 5% or 10% of highest-scoring holdings equal to that of an *average* holding (for the crop in question). The results of such an exercise are shown in Fig. 7.8. As can be seen, the pay-off of tackling the worst 5% or 10% polluters varies widely from crop to crop. These differences are related to the variation found within each individual crop (intra-crop variation; *cf.* Figure 7.7). The greatest benefits to be gained from controlling the top 5% are in winter wheat (61%), followed by spring barley and sugarbeet. The gains are far smaller (9–10%) for such crops as apples, pears and tulips. If the top 10% bracket is tackled, a reduction in environmental impact of almost 80% can be achieved for spring barley and 72% for winter wheat; for apples this is only 11%.

Factoring the respective national crop acreages into the equation, we can now gauge the effectiveness of controlling the worst 5% or 10% of polluters in reducing the overall environmental impact of agriculture in The Netherlands. These calculations show that controlling the worst 5% of offenders for each specific crop would lead to a 23% reduction in aggregate environmental impact, while a 33% improvement could be achieved at the national level if the worst 10% were tackled. These greatest reductions can be secured by tackling arable farms (almost 70% of benefits in both cases) and in particular potato growers (ware, industrial and seed). Here, the large scale of operations is the major contributing factor.

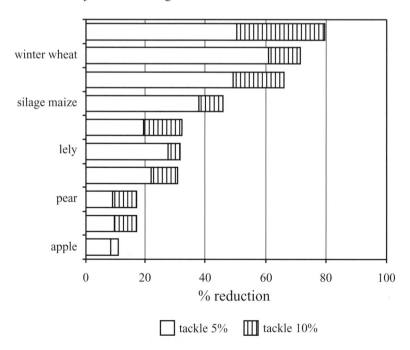

Fig. 7.8   Reduction in environmental impact to be secured by controlling the worst 5% or 10% polluters.

In a similar fashion farm-scale computer runs can also be carried out for other forms of preventative policy, such as measures to reduce the pollution of water-courses. In this context a number of studies have shown that pesticide drift to surface waters can be cut dramatically by implementing buffer zones, particularly if these are combined with use of low-drift spray nozzles or other 'smarter' application tech-niques (van der Zande *et al.* 1995; de Snoo & de Wit 1998). A 95% reduction in pesticide drift to surface waters, a feasible figure, would reduce the overall environ-mental burden associated with pesticide use in The Netherlands by 87% (de Snoo & de Jong 1999).

Finally, an important question at the present juncture is whether and to what extent the preventative strategies outlined above can be implemented and enforced. This would require the full cooperation of arable farmers and other growers, for it is they who make the ultimate decisions about whether or not to treat their crops, which product to use and in what dosage, and what equipment to employ for the task. They are also the ones doing the actual spraying and in a position to accomplish substantial environmental benefits if they make the appropriate choices.

In this context there is a role not only for government but certainly also for the private sector: farmers' organisations, educators, pesticide merchants and other parties involved in the agro-production chain, such as supermarket chains and whole-salers (*cf.* van der Grijp in this volume). After all, no one wishes to see a policeman stationed at every farmyard. Market incentives such as voluntary 'green' certification schemes for farm produce are a superior option (see Udo de Haes & de Snoo 1996; 1997; Oskam *et al.* 1998). With the computer facilities now available, encouragement might also be given to the benchmarking of pesticide usage and associated environ-mental impact to facilitate inter-holding comparison.

References

Buurma, J.S. (1997) *Oorzaken van verschillen in middelenverbruik tussen bedrijven. Schurftbe-strijding in appelen.* LEI-DLO publication 4.143, The Hague.

CBS [Centraal Bureau voor de Statistiek] (1994) *Gewasbescherming in de land- en tuinbouw, 1992. Chemische, mechanische en biologische bestrijding.* Sdu/CBS, The Hague.

CBS [Centraal Bureau voor de Statistiek] (1997) *Gewasbescherming in de land- en tuinbouw, 1995. Chemische, mechanische en biologische bestrijding.* CBS, Voorburg.

CLM [Centrum voor Landbouw en Milieu] (1997) *Milieumeetlat 1997, werkboek, milieumeetlat voor bestrijdingsmiddelen.* CLM, Utrecht.

EC [European Commission] (1996) Possibilities for future EU environmental policy on plant protection products. Sub-report: Regional analysis of use patterns of plant protection prod-ucts in six EU countries. PES – A/Phase 2. A comparison of agrochemical use on potatoes in four regions in Europe.

EC [European Commission] (1997) Possibilities for future EU environmental policy on plant protection products. Synthesis Report B.

IKC [Informatie en Kennis Centrum] (1996) *Achtergronden van de milieumeetlat voor bestrijdings-middelen.* IKC (MJP-G unit), Wageningen.

de Jong, F.M.W. & de Snoo, G.R. (2002) A comparison of the environmental impact of pesticide use in integrated and conventional potato cultivation in The Netherlands. *Agriculture, Ecosystem and Environment*, 91, 5–13.

de Jong, F.M.W, de Snoo, G.R. & Loorij, T.P.J. (2001) Trends of pesticide use in The Netherlands. Proceedings of the International Symposium on Crop Protection, Gent.

Oskam, A.J., Vijftigschild, R.A.N. & Graveland, C. (1998) *Additional EU Policy Instruments for Plant Protection Products*. Wageningen Pers, Wageningen.

Reus, J., Leendertse, P., Bockstaller, C., Fomsgaard, I. *et al.* (2002) Comparison and evaluation of eight pesticide environmental risk indicators developed in Europe and recommendations for future use. *Agriculture, Ecosystems and Environment*, 90 (2), 177–87.

de Snoo, G.R. & de Jong, F.M.W. (1999) *Bestrijdingsmiddelen en milieu*. Uitgeverij Jan van Arkel, Utrecht.

de Snoo, G.R., de Jong, F.M.W., van der Poll, R.J., Janzen, D.E., van der Veen, L.J. & Schuemie, M.P. (1997) Variation of pesticide use among farmers in Drenthe: A starting point for environmental protection. *Mededelingen van de Faculteit Landbouwwetenschappen, Ryksuniversiteit Gent* 62/2a, 199–212.

de Snoo, G.R. & de Wit, P.J. (1998) Buffer zones for reducing pesticide drift to ditches and risks to aquatic organisms. *Ecotoxicology and Environmenal Safety*, 41, 112–18.

Solomon, K. (1999) Probabilistic Risk Assessment of Agrochemicals. BBA Workshop on *Risk Assessment and Risk Mitigation Measures in the context of the Authorization of Plant Protection Products*. 27–29 September 1999. Braunschweig.

Udo de Haes, H.A. & de Snoo, G.R. (1996) Environmental certification. Companies and products: two vehicles for a life cycle approach? *International Journal of Life Cycle Assessment* 1 (3), 168–170.

Udo de Haes, H.A. & de Snoo, G.R. (1997) The agro-production chain. Environmental management in the agricultural production-consumption chain. *International Journal of Life Cycle Assessment*, 2 (1), 33–8.

van der Zande, J.C, Holterman, H.J. & Huijsmans, J.F.M. (1995) *Driftbeperking bij de toediening van gewasbeschermingsmiddelen. Evaluatie van de technische mogelijkheden met een drift model*. IMAG-DLO, Wageningen.

Chapter 8

Assessing the Environmental Performance of Agriculture

Pesticide Use, Risk and Management Indicators

Kevin Parris and Yukio Yokoi

Introduction

A central theme of this book is to analyse the effectiveness of pesticide management and related policies in limiting the negative externalities to human health and the environment from the use of pesticides. This chapter helps address this challenge by examining recent efforts across OECD (Organisation for Economic Cooperation and Development) countries[1] over the past fifteen years to monitor changes in agricultural pesticide use and their risks to human health and the environment, and to track how farmers are changing their pest management practices.

This chapter sets out to describe indicator methodologies being developed to help measure changes in agricultural pesticide use, risks and pest management, focusing on recent trends in these indicators for OECD countries. The chapter first outlines the policy and environmental context to the pesticide use, risk and management issues. The third section defines the indicators and their methods of calculation, followed by a description of recent indicator trends, and the chapter concludes by examining future challenges to further research in this area.

The policy and environmental context

Policy context

Policymakers need to address a range of human health and environmental issues associated with the external costs of pesticide use, including:

- the exposure of farm workers and the public in the vicinity of where pesticides are applied;
- consumer exposure to pesticide residues in food;
- potential human health risks that are not well understood, for example, hormonal effects;
- contamination of ground- and surface water used for drinking by both humans and livestock; and

- environmental impacts on terrestrial and aquatic habitats, such as risks to non-targeted organisms and wildlife.

OECD countries have, broadly, used three main types of policy approaches to address the human health and environmental issues linked to pesticide use: economic instruments, regulatory measures, and information and voluntary approaches.

Economic instruments

While pesticide use is sometimes subsidised (as in Turkey) and in other cases taxed (e.g. Denmark, Norway, Sweden),[2] farmers usually pay the market price, although they do not always pay the 'full' or 'social' cost of production. This is because the market price of pesticides does not fully reflect the external costs resulting from their impact on the environment and human health (Pearce & Tinch 1998).

Regulatory measures

A key aspect of pesticide policies in OECD countries is the regulatory system that assesses pesticides before they can be approved for sale and use (*cf.* Vogelezang-Stoute in this volume). The registration process is to ensure that pesticides do not pose unacceptable human health and environmental risks above nationally agreed thresholds. Moreover, most OECD countries have legal standards with respect to maximum permissible residue levels both for individual pesticides and for total pesticide substances in food and drinking water.[3] Even so, uncertainties remain concerning pesticides risks, for example the so-called 'cocktail effect', that is, the risk associated with combinations of pesticide residues in food and water.

Information and voluntary approaches

Some countries, for example Denmark, The Netherlands and Sweden, have set targets to reduce the total quantity of agricultural pesticides used over a given period. Many of the targets that were originally set in terms of tonnes of active ingredient are now being revised to focus on the reduction in pesticide risks and improving pest management practices, particularly encouraging the uptake by farmers of integrated pest management through farm extension, advisory and information schemes.

There are also a number of multilateral efforts to reduce the harmful health and environmental effects from pesticide use. The aim of the European Union's Fifth Environment Action Programme is to achieve a significant reduction in pesticide use per unit of agricultural land. Thus, European countries participating in the North Sea Treaty (1983) have commitments to reduce emissions of certain pesticides. Among other things, the Treaty has called for countries to ban or restrict 18 pesticides and reduce by 50% emissions of 36 other pesticides near marine waters. A number of OECD countries bordering the Baltic Sea have also made commitments to reduce emissions of pesticides under the Baltic Sea Treaty (1974).

Canada and the United States have projects aimed at preventing pesticide contamination of the Great Lakes. Under the North American Free Trade Agreement Technical Working Group on Pesticides, there is a commitment to work together towards a single North American market for pesticides, while maintaining current high levels of protection of public health and the environment, and supporting the principles of sustainable pest management.

Internationally, the FAO/WHO CODEX Commission has established maximum residue limits on pesticide residues in fruit and vegetables (Gebbie 1998). Furthermore, it was agreed, under the Montreal Protocol on Substances that Deplete the Ozone Layer (1987), that methyl bromide (mainly used as a soil fumigant by agriculture) should be phased out by 2005, with possible exemption for critical agricultural uses (EEA 1995; Oberthur 1997; UNEP 2000).

Environmental context

The quantity of pesticide applied by farmers depends on the level of pest and disease pressure, climatic conditions, the type of crop and its resistance to pests and disease, the efficiency of pest management practices, and the influence of economic and policy factors. Moreover, the amount of pesticides that leach into soil and water or evaporate into the air depends on site-specific conditions, such as soil properties and temperature, drainage, type of crop, climate, and application method, time and frequency. The risks posed by different pesticides vary greatly depending on their inherent toxicity (or hazard) and exposure that can occur based on the pesticide's mobility and persistence in the environment and the method and quantity applied.

The *mobility* of pesticides in the environment is mainly determined by the type of pesticide, the rate of pesticide uptake by different crops, topography and soil type, and the climatic conditions where the pesticides are applied. Some of the pesticides applied can evaporate and possibly photo-decompose. The fate and mobility of remaining pesticides depends on the organic content of soil, and soil erosion, leaching and run-off rates. The last are in direct relation to the climatic conditions of a specific drainage basin. Estimates vary widely as to the quantity of pesticides applied that actually reach the target pests, from less than 1% to 75%, with the remainder lost to the environment through soil run-off, erosion, leaching and vaporisation into the atmosphere.

The *persistence* of pesticide residues in the environment and human food chain may vary from a few weeks to 30 years. Despite the ban on DDT in most OECD countries since the mid-1970s, for example, residues of this pesticide compound are still detectable in some aquatic environments, such as in the United States (USGS 1999). Research also shows that approximately 10% of all herbicides have a persistency in the soil that may adversely affect the yield of crops following those to which the herbicides were first applied (EEA 1995).

Pesticides vary in their *degree of toxicity* depending on the type and concentration of their active ingredients (the chemicals actually controlling or killing the intended pest, weed or disease). When pesticides that are less toxic are used, environmental damage may decrease despite increases in pesticide use. Moreover, the sensitivity of wildlife to toxic contamination varies both with specific pesticides and with wildlife species. In the United Kingdom, for example, trends in pesticide use show an overall decline in use of products that are acutely toxic to mammals, but an increase in pesticides with high acute toxicity to aquatic organisms (Department of the Environment 1996).

The quantification of *human health risks* from exposure to pesticides in foodstuffs is complex, while some uncertainties remain concerning the validity of extrapolating to human health from laboratory tests of pesticide contaminants on animals. In addition, there is the problem of separating out the effects of pesticides from the many other influences on human health, such as the composition of the diet including tobacco and alcohol, and also age, gender and ethnic background. However, many OECD countries regularly sample and test food products for evidence of pesticide residues, with detection methods improving rapidly.

Similarly, the quantification of *risks to terrestrial flora and fauna* from pesticide use is also complex. Pesticides can accumulate in food chains with consequent indirect impacts along the food chain, while they may directly eradicate, remove or reduce food sources for birds and mammals (Rayment *et al.* 1998). In aquatic environments the leaching of pesticides into rivers, lakes and coastal waters is known to cause damage to aquatic biodiversity.

Development of *pest resistance* to pesticides is mainly an economic problem, though not a health or environmental concern unless it leads to the use of more hazardous substitute pesticides or to increased damage to agricultural crops, or both. In the United States, for example, 183 insect pests are resistant to one or more insecticides, and 18 weed species are resistant to herbicides (USDA 1997). The use of genetically modified plants to overcome such problems might be an area of considerable potential, although there is a major international research effort under way to examine the environmental and human health effects of genetic engineering.

It is estimated that methyl bromide accounts for 5–10% of the global loss of stratospheric ozone, and may be responsible for around 20% of the Antarctic ozone depletion (Mano & Andreae 1994). Developed countries account for about 80% of methyl bromide use world-wide. The main sources of methyl bromide are vehicle exhaust (from vehicles using leaded petrol), emissions from plankton in the oceans, burning biomass (including grassland and forest fires) and agricultural pesticide use. Methyl bromide is used as a soil fumigant, and it is estimated that this accounts for 90% of total use in the European Union (EUROSTAT 1999). According to research by Mano & Andreae (1994), agricultural pesticide use as a source of methyl bromide accounts for 25–60% of total annual global emissions. Grassland and forest fires also provide a major contribution of around 30% to the annual stratospheric bromine budget.[4]

Measuring trends in agricultural pesticide use, risks and pest management

The basic long-term challenge for agriculture is to produce food and industrial crops efficiently, profitably and safely, and to meet a growing world demand without degrading natural resources and the environment. In order to respond to the challenge and develop better policies, policymakers need agri-environmental indicators, which can help to monitor the environmental effects of agriculture and provide a tool for policy analysis. Pesticide indicators can provide a useful tool for the evaluation of domestic policies and international obligations related to pesticide use in agriculture. Such indicators can also convey a general idea about trends in pesticide use, risk, and management, and the impact of pesticides on human health and the environment.

Many OECD countries are developing three main types of indicators, covering the (a) use, (b) risks and (c) management of pesticides. Fig. 8.1 provides a simplified overview of the various linkages between pesticide use, risk, management and other agri-environmental indicators. Pesticide use is influenced by whole farm management practices adopted by farmers; for example, the adoption of organic farming systems will lower pesticide use. Also the use of specific pest management practices, such as integrated pest management, will also affect the use and associated risks from pesticide use.

The impact of pesticide use on human health, concern the direct effects in terms of exposure to farm workers and the public in the vicinity of spraying. There are also indirect effects through pesticide residues in food and water consumption, with related concerns such as pesticide poisoning, cancer and endocrine disruption. The risks to the environment from agricultural pesticide use concern impacts on terrestrial and aquatic flora and fauna, toxic contamination of soils, and the links between methyl bromide emissions and ozone depletion.

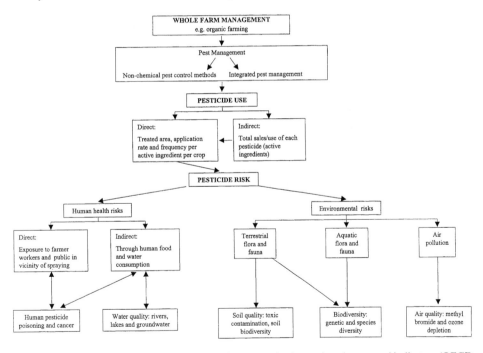

Fig. 8.1   Linkages between pesticide use, risk indicators, and other agri-environmental indicators (OECD 2001).

At present OECD countries' research on developing pesticide indicators has mainly concentrated on the indirect change in total sales of pesticides, and on pesticide risks to the aquatic environment, although many countries are beginning to develop risk indicators to cover human health and terrestrial environmental risks.[5] Pesticide use indicators are simpler and more straightforward because they deal with just one type of information rather than combining different types. However, because OECD country policies aim ultimately to reduce risks and not merely pesticide use, it is important to develop the more complex risk indicators that could help measure the effectiveness of these policies.

Pesticide use

Indicator definition and method of calculation
The indicator of pesticide use shows trends over time based on pesticide sales or use data, or both, and measured in tonnes of active ingredients. The three-year average covering 1985–1987 is used here as the base year to reduce the impact of extreme values and also to reflect changes since the period when many OECD countries began the process of agricultural policy reform to lower support to agriculture. The pesticide use indicator is calculated as:

$$\frac{(\text{Quantity of pesticides used in year } t)}{(\text{Average quantity of pesticides used in } 1985-87)} \times 100$$

The indicators of pesticide use track trends over time in the overall quantity of pesticide used. Although the term 'pesticide use' is used here, only a few OECD countries have data on actual use and the term generally refers to data on pesticide sales, which is often used as a proxy for pesticide use. For most countries total pesticide use data includes four main sub-categories: herbicides (defoliants and desiccants); insecticides (acaricides, molluscicides, nematocides and mineral oils); fungicides (bactericides and seed treatments); and other pesticides (fumigants, rodenticides, anti-coagulants, growth regulators and animal repellents).

National indicators of pesticide use serve various purposes, such as to evaluate trends in pesticide use over time as a crude proxy for potential reduction in risks, and to reveal possible improvements in pesticide use efficiency if crop production is increasing more rapidly than use. They can also determine whether lower than recommended rates of pesticide use are effective, and help evaluate whether the use of integrated pest management and other specific farm management practices and policy actions reduce pesticide use.

Recent trends
Several key points emerge from the recent trends in pesticide use data shown in Fig. 8.2. Overall the trend in pesticide use over the last decade has remained constant or declined for most OECD countries, although pesticide use increased for a number of them. For those countries where pesticide use has increased, this has, in general, been in response to an expansion in crop production, as illustrated by the examples of Belgium, Greece, Ireland and Korea.

A significant reduction in pesticide use has occurred in the Czech Republic, Hungary and Poland, which to a large extent can be explained by their transition to a market economy since the early 1990s (Fig. 8.2). The sharp reduction in pesticide use in these countries has been mainly due to the collapse in agricultural support levels, the elimination of subsidies for pesticides, and increasing debt levels in the farm sector limiting farmers' ability to purchase such inputs (OECD 1998).

Significant reductions in pesticide use, by 30% or more over the past ten years, are also observed in countries that have set targets to reduce the use of pesticides. Examples include Denmark, Finland, The Netherlands, Norway and Sweden. The reduction has also been linked to the increasing area of crops under organic farming and subject to integrated pest management and other pesticide reduction practices, for example in Italy, Spain and Switzerland.[6]

The expansion in the area under organic farming is also acting to reduce pesticide use in some countries, for example in Austria, Finland, Germany, Italy, Sweden and Switzerland (Fig. 8.4). Decreasing pesticide use in the United Kingdom has been due, in particular, to the introduction of new herbicides with lower recommended doses (MAFF 2000).

In Japan the reduction in pesticide use has closely reflected the declining trend in crop production, in particular the decrease in rice production, Japan's major crop (Fig. 8.2). In New Zealand pesticide use rose steadily from 1985 reaching a peak in 1996. According to a recent study, however, usage declined by about 10% in 1998, largely reflecting the drop in crop production during that year (Holland & Rahman 1999).

From the early 1980s up to the 1990s pesticide use decreased in the United States, as commodity prices fell and large areas of agricultural land were taken out of production under government programmes (Table 8.1). Since 1990 US pesticide usage has fluctuated with changes in planted area, infestation levels, adoption of new products and other factors, including the increasing adoption of integrated pest management practices by farmers (see Fernandez-Cornejo & Jans 1999; and USDA 1997).

Indicator interpretation and limitations

The definition and coverage of pesticide use data vary across OECD countries, which limits the use of the indicator as a comparative index. Only a few countries have data on actual pesticide use, but nearly all OECD countries report data on pesticide sales, which can be used as a proxy for pesticide use, although ideally it should be supported by representative samples of the use data. For some countries, series are either incomplete, especially over recent years, or do not exist.[7] The OECD, in cooperation with EUROSTAT, is beginning a process to help improve the collection of pesticide use data, see, for example, (OECD 1999c). A further difficulty is to identify pesticide use specific to agriculture, net of uses for forestry, gardens, golf courses and so on, and the quantity of pesticides used for specific crops and pasture, although some limited data are available on the latter.

Studies in a few OECD countries suggest that, at least over the short term, there is in some cases a correlation between trends in pesticide use and environmental risks: as use declines, risks also decrease. However, some caution is required in making this link, for a number of reasons. A change in pesticide use may not be equivalent to a change in the associated risks because of the continually changing pesticide market and the great variance in risks posed by different products.

Changes in the herbicide market seen in the 1980s provide a good illustration. During this period, new herbicide products came onto the market that were much

Table 8.1 Total use of agricultural pesticides in OECD countries 1985–1997 (tonnes of active ingredients) (OECD Environmental Data Compendium 1999d; EUROSTAT 1999; Holland & Rahman 1999).

	1985	1986	1987	1988	1989	1990	1991	1992	1993	1994	1995	1996	1997
Australia	—	—	—	—	—	—	—	119.654	—	—	—	—	—
Austria	5.270	6.069	—	—	4.615	4.246	4.487	3.897	3.984	3.619	3.402	3.565	3.690
Belgium	8.748	8.748	8.923	9.535	9.885	9.973	9.623	10.060	9.885	9.510	10.536	9.976	8.619
Canada	39.259	32.968	33.883	35.529	—	33.964	—	—	—	29.206	—	—	—
Czech Republic	—	—	—	—	11.217	8.920	6.361	4.817	3.645	3.680	3.783	3.908	3.889
Denmark	6.863	6.085	5.485	5.253	5.795	5.650	4.628	4.566	4.103	3.919	4.809	3.669	3.675
Finland	1.964	1.933	1.988	1.923	2.258	2.037	1.734	1.410	1.260	1.297	1.054	933.000	1.016
France	98.027	99.697	92.966	99.167	100.433	97.701	103.434	84.709	91.953	89.515	84.006	97.890	109.792
Germany	—	—	—	—	—	—	—	—	—	—	—	—	—
Greece	—	7.346	6.510	6.754	8.151	—	7.860	8.567	8.583	9.973	8.525	9.870	9.034
Hungary	26.342	31.818	26.918	25.341	35.438	25.501	16.129	11.541	10.195	9.560	7.696	—	—
Iceland	—	—	—	—	—	—	—	—	—	—	—	—	—
Ireland	—	—	—	1.812	1.899	1.745	1.915	1.942	2.169	2.160	2.255	1.741	2.325
Italy	99.579	—	99.100	100.579	91.070	91.680	58.123	58.848	54.928	46.678	48.490	48.050	—
Japan	—	97.550	95.886	94.096	93.347	92.608	88.014	86.718	87.270	87.598	86.331	83.678	84.541
Korea	—	21.322	23.229	21.967	23.280	25.082	27.476	26.718	25.999	26.282	25.834	24.541	24.814
Luxembourg	—	—	—	—	—	—	253	—	—	—	—	—	—
Mexico	—	—	—	—	—	—	—	—	36.000	—	—	—	—
Netherlands	21.002	21.632	18.088	18.172	19.146	18.835	17.206	15.951	11.761	11.169	10.923	10.338	10.397
New Zealand	3.690	—	—	3.732	—	—	—	—	—	3.757	—	3.752	—
Norway	1.529	1.514	1.323	1.194	1.035	1.184	771	781	765	862	931	706	754
Poland	12.398	14.479	18.444	23.377	20.620	7.548	5.217	6.755	6.791	7.335	6.962	9.420	9.501
Portugal[1]	—	—	—	—	—	—	9.355	6.117	8.984	9.581	11.818	12.457	12.751
Spain	—	39.134	44.050	47.751	46.534	39.562	39.147	31.839	29.408	31.243	27.852	33.236	34.023
Sweden	3.660	5.585	2.409	2.865	2.423	2.344	1.837	1.512	1.464	1.961	1.224	1.528	1.609
Switzerland	—	—	—	2.456	2.464	2.283	2.056	2.022	1.936	1.921	1.827	1.747	—
Turkey	—	—	—	—	—	—	—	—	—	—	—	—	—
United Kingdom	40.826	40.759	40.719	32.985	32.643	35.858	35.364	31.696	32.400	33.945	12.500	13.976	15.575
United States	390.894	372.280	369.556	383.630	365.924	378.636	370.918	380.564	367.863	33.945	33.774	35.523	35.432

— not available
[1]Sulphur is responsible for about 50 % of the total indicated values.

Change in tonnes of active ingredients   Tonnes of active ingredients[5]

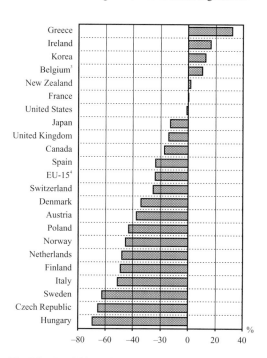

	1985–87	1995–97
Greece	6,928	9,143
Ireland	1,812	2,107
Korea	22,276	25,063
Belgium	8,806	9,710
New Zealand	3,690	3,752
France	96,897	97,229
United States	377,577	373,115
Japan	97,672	84,850
United Kingdom	40,768	34,910
Canada	35,370	29,206
Spain	41,592	31,704
EU-15	333,804	253,684
Switzerland	2,546	1,832
Denmark	6,144	4,051
Austria	5,670	3,552
Poland	15,107	8,628
Norway	1,455	797
Netherlands	20,241	10,553
Finland	1,962	1,001
Italy	99,100	48,270
Sweden	3,885	1,454
Czech Republic	11,217	3,860
Hungary	28,359	8,628

Fig. 8.2 Pesticide use in agriculture in OECD countries, 1985/1987[1] to 1995/1997[2] (OECD Environmental Data Compendium 1999d; EUROSTAT 1999; Holland & Rahman 1999).

Notes: Some caution is required in comparing trends across countries because of differences in data definitions and coverage.
1. Data for 1985–87 average cover: 1986–87 average for Greece, Korea, and Spain; 1985 for New Zealand; 1985–86 average for Austria; 1987 for Italy; 1988 for Ireland and Switzerland; and 1989 for the Czech Republic.
2. Data for 1995–97 average cover: 1994–95 average for Hungary; 1994–96 average for Switzerland; 1995–96 average for Italy; 1991–93 average for the United States; 1994 for Canada; and 1997 for New Zealand.
3. Includes Luxembourg.
4. Excludes Germany and Portugal.
5. The following countries are not included in the figure: Australia, Germany, Iceland and Mexico (time series are not available); Portugal (data are only available from 1991); Slovak Republic (became an OECD member in 2000); and Turkey (data are only available from 1993).

more biologically active than their predecessors and were therefore used in much smaller quantities. Pesticide use indicators for this period would show a substantial reduction in herbicide use. By contrast, risk indicators might show no change, or perhaps even an increase, in the environmental and human health risks associated with herbicide use. In addition, the greater use by farmers of pesticides which carry a lower risk to humans and the environment, because they are more narrowly targeted or degrade more rapidly, might also not reveal any change in overall pesticide use trends, and possibly even an increase.

There are an enormous number of pesticide products available for farmers to use. For example, over 700 pesticide products (active ingredients) are marketed in the European Union, each of which poses unique environmental and health risks. With respect to risks to water quality, however, a recent French study found that, while more than a hundred products are detected at variable concentrations and frequencies in water, most of the water pollution from pesticides in France is caused by about ten products. These are mainly herbicides belonging to the triazine family (IFEN 1998).

Care is also required when comparing trends in pesticide use across countries, because of differences in climatic conditions and farming systems, which affect the composition and level of usage. Variability of climatic conditions, especially moisture, may markedly alter pesticide use. Warmer conditions generally require higher levels of use than colder conditions to maintain agricultural productivity. In the US, for example, the sweet corn crop is typically treated with insecticides seven to fourteen times annually in southern, warmer regions of the country, compared with only two to four treatments in the northern, colder regions. In the southern states over 20% of the rice acreage is treated with fungicides for rice blast disease, which is not a problem in California where no fungicides are used (OECD 1997). However, not all pesticide use increases with warmer weather, herbicide use being an example.

Changes in cropping and rotation systems, tillage practices, the uptake of integrated pest management practices, the use of precision farming technology and the expansion of organic farming can also affect agricultural pesticide use. The change in agricultural cropping systems from arable and permanent crops to forage, for example, will usually lead to a significant reduction in pesticide use. It is for this reason that the commonly used indicator showing pesticide use per hectare of total agricultural land can be misleading when compared across countries.

The usefulness of pesticide use indicators can be improved by linking them to pesticide risk indicators and to other indicators, particularly those covering soil and water quality and farm pest management. For example, there is some evidence that moving from intensive farm practices to integrated pest management (IPM) and organic farming systems may achieve a considerable reduction in pesticide use, while maintaining the economic viability of the system (OECD 1997). On the other hand, maintaining winter green cover to limit nutrient losses from agricultural land, for example, can require the additional use of pesticides.

Pesticide risk

Pesticide risk indicators show trends in risks over time by combining information on pesticide hazard and exposure with information on pesticide use. The OECD has developed three models that can be used to calculate indicators of pesticide risk to aquatic organisms (work on indicators for other risk areas, i.e. terrestrial and human health risk, is under way). The three models are designed to produce aggregate risk trends at a national level; however, they can also be used to calculate risk trends for smaller areas. In addition, all three methods can be used to calculate trends for short-term (acute) and long-term (chronic) aquatic risks, and at different levels of aggregation: for one, several or all pesticides; one, several or all crops; and one, several or all aquatic organisms.[8]

A growing number of OECD countries have also developed pesticide hazard or risk indicators. In general, these indicators are intended to help measure progress in meeting the goals of national risk reduction programmes. Despite the high interest in pesticide risk indicators, and the considerable research on them in recent years, there is no consensus on a single methodology that all countries could use. This is partly because individual governments wish to use indicators for different purposes (depending on the focus of their risk reduction programme, for example), and partly because risk indicator models are difficult to design, where risks are influenced by a multitude of factors that vary within and across countries. The OECD is, therefore, focusing initially on the development and testing of different pesticide risk indicator models rather than on reporting risk trends in different countries.

Indicator definition and method of calculation

Pesticide risk indicators show trends in risk over time by combining information on pesticide toxicity and exposure with information on pesticide. Three methods being developed by OECD are intended to represent the range of approaches that could be used to calculate aquatic risk indicators. In particular, they draw on characteristics of the indicator models developed by Denmark, France, Germany, The Netherlands and Sweden. The indicators share some basic features, including that:

- they use identical data on pesticide toxicity and similar data on other pesticide characteristics, such as fate and behaviour in the environment; and
- they have the same basic structure as follows:

$$\text{Pesticide risk} = \frac{\text{exposure}}{\text{toxicity}} \times \text{area treated}$$

where exposure equals the level of pesticide estimated to occur in water bodies adjacent to farm fields; toxicity is the level that would be harmful to aquatic organisms, e.g. the level that is lethal to 50% of the organisms exposed; and area treated is the number of hectares on which the pesticide was used.

The way the indicators differ is in how they calculate exposure. For this, they use different combinations of the two basic approaches used in other national risk indicator work, namely scoring and the use of a mechanistic model. The *scoring approach* converts data relevant to exposure into scores that reflect their general contribution to exposure, then combines the scores in ways that give appropriate weight to each variable. The *mechanistic approach* combines the actual data values through a series of mathematical equations that mirror scientific understanding of environmental processes that contribute to exposure.

The three methods, which OECD has been developing on the basis of the scoring and mechanistic approaches, are:

(1) ratio of exposure to toxicity (REXTOX): based entirely on the mechanistic approach;
(2) additive scoring (ADSCOR): uses a simple scoring system but includes some original (unscored) variables; and
(3) synergistic scoring (SYSCOR): uses a more complex scoring system and some original (unscored) variables.

REXTOX is calculated as follows:

$$\text{REXTOX}_{\text{short-term}} = \frac{\text{ADR} \times (\text{LOSS} / \text{Water depth}) \times \text{Water index} \times \text{AFT} \times \text{BAT}}{\text{short-term toxicity}}$$

$$\text{REXTOX}_{\text{long-term}} =$$
$$\frac{\text{ADR} \times (\text{LOSS} / \text{Water depth}) \times \text{Water index} \times \text{AFT} \times \text{LTF} \times \text{BAT}}{\text{long-term toxicity}}$$

where ADR is the actual dose rate; LOSS is the amount of pesticide that escapes into water bodies due to spray-drift and run-off, taking account of the crop grown, the pesticide application method, the presence and size of untreated buffer zones, etc.; Water depth is the depth of water bodies (e.g. rivers, lakes); Water index is the proportion of the treated area bordered by surface water; AFT is the average frequency of treatments; BAT is the basic area treated; LTF is the long-term factor (ratio of concentration of the pesticide concerned over a certain period and the initial concentration, with the default value of 21 days); short-term toxicity: for fish, is 50% lethal concentration (LC50) over 96 hours; for Daphnia, 50% effect concentration (EC50) over 48 hours; and for algae, 50% effect concentration (EC50) over 96 hours; long-term toxicity: for fish, Daphnia and algae, no observable effect concentration (NOEC) over 21 days.

ADSCOR is calculated as follows:

$$\text{ADSCOR}_{\text{short-term}} = \frac{(\text{short-term exposure score} + 1) \times \text{BAT}}{\text{short-term toxicity}}$$

$$\text{ADSCOR}_{\text{long-term}} = \frac{(\text{long-term exposure score}) \times \text{BAT}}{\text{long-term toxicity}}$$

where short-term exposure score is the sum of five scores for average actual dose rate, frequency of treatments per harvesting season, method of application, spray-drift buffer zone, run-off buffer zone, and water index; long-term exposure score is the short-term exposure score above, plus the sum of six scores for half-life (DT50) in water, photolysis in water, LogKow, half-life (DT50) in soil, Koc, and water index; where Photolysis is chemical decomposition induced by light or other energy; LogKow is the standard system used often in the assessment of environmental fate and transport for organic chemicals, and is a measurement of how a chemical is distributed at equilibrium between octanol and water; and Koc is a measure of a material's tendency to adsorb soil particles, measured as the ratio of the chemical adsorbed per unit weight of organic carbon in the soil or sediment to the concentration of the chemical in solution at equilibrium, with high Koc values indicating a tendency for the material to be adsorbed by soil particles rather than remain dissolved in the soil solution.

SYSCOR is calculated as follows:[9]

$$\text{SYSCOR}_{\text{short-term}} = \frac{\text{exposure score (including area treated factor)}}{\text{short-term toxicity}}$$

where exposure score is the combination of 9 scores for cumulative area treated, actual dose rate, method of application, users' training level, water index, solubility in water, half-life (DT50) in water, half-life (DT50) in soil, and LogKd; where LogKd is the soil–water adsorption coefficient, calculated by using measurements of pesticide distribution between soil and water.

Simplified formula for the three indicators are being considered by OECD countries, and will be tested and their results compared with those of the three indicators described here. Their formulae are:

- REXTOX = tonnes applied / toxicity / buffer;
- ADSCOR = area treated * buffer / toxicity; and
- SYSCOR = SCORE (area treated, buffer) / toxicity

Recent trends
Initial testing of the three methods for aquatic risk (REXTOX, ADSCOR and SYSCOR) was completed using pesticide use data on arable crops and orchards in England and Wales. The risk trends produced by the three indicators for total pesticide use on arable crops between 1977 and 1996 are shown in Fig. 8.3. The results show that different indicator methods can produce different pesticide risk trends, even when using the same data set.

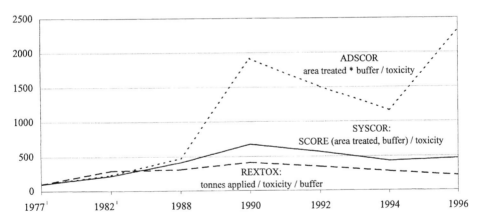

Fig. 8.3 Aquatic risk from pesticide use on arable crops: England and Wales 1977–1996 (Index 1977 = 100) (MAFF, UK, unpublished).

The relative contribution of single pesticides to the total risk was also analysed in the indicator trial. It was found out that the use of the herbicide cypermethrin contributed most to the risk trends produced by REXTOX and ADSCOR, and also figured importantly in SYSCOR. The trends diverge after 1988 because of the different ways the indicators deal with pesticide dose rate and untreated buffer zones bordering water bodies, which have been required for cypermethrin in England and Wales since 1992.

The next stage of the OECD work on pesticide risk indicators is a 'pilot project' in which OECD countries are using REXTOX, ADSCOR and SYSCOR with their own national pesticide data. The purpose is to see how easy the different methods are to use, how the results compare, and how closely the trends they produce correspond to expected risk trends. OECD countries that have developed separate aquatic risk indicator methods are including these in the project as well, to enable comparison of an even broader range of indicator approaches.

Indicator interpretation and limitations
The OECD project has identified the strength and weakness of the three methods for pesticide indicators, which are summarised below.

REXTOX
- Using precise endpoint values rather than scores, REXTOX is the most responsive of the three indicators to changes in input values. It can also be easily adapted to different regional conditions, such as weather, soil and physical features like slope.
- REXTOX is relatively objective and transparent. By using direct input values and models to calculate pesticide levels in water bodies, which are similar to the ones used for risk assessment, REXTOX minimises reliance on expert judgement to set scores, weight variables, and so forth. This objectivity is only relative, however, because expert judgement was required to establish the indicator and to choose which models to incorporate.
- The precise estimates produced by REXTOX's exposure models rely on various assumptions about exposure processes that may or may not be correct. The indicator results may thus imply a 'false precision'.
- REXTOX is quite complex. Although scientists and risk assessors may consider it transparent and clear, its formulae may be difficult for others to understand.

ADSCOR
- ADSCOR's basic structure and equation are easy to understand, even by those without technical expertise. ADSCOR is also relatively easy to modify, if a user wants to add new parameters or delete existing ones. Such changes require a consideration of the relative risk contribution of any added parameters, but do not involve complicated mathematical models.
- By expressing risk factors in a qualitative way (low, medium, high), ADSCOR may be easier to grasp than, for example, a precise value for water solubility. In addition, the use of scores makes ADSCOR less demanding on data needs by including ranges rather than exact values for some parameters (e.g. $DT_{50} > 60$ days).
- Converting the input values into scores results in a loss of precision and 'sensitivity' to minor changes in the values. Scoring indicators can over- or underestimate such changes depending on where the values fall in relation to the 'breakpoints' between the scores. Moreover, assigning scores and weighting the different variables is subjective (based on expert judgement) and dependent upon local conditions that affect pesticide risk.
- ADSCOR and other scoring indicators may require some modification before actual use. The equation to combine the scores will remain constant, but each user

will need to review – and in many cases re-establish – scores and their classification categories.

SYSCOR
- As a scoring indicator, SYSCOR shares many of the advantages and disadvantages identified for ADSCOR. However, with its synergistic scoring system, SYSCOR incorporates better than most scoring indicators scientific understanding of the interactions among environmental fate and exposure processes. The disadvantage is that the system is complicated and not fully transparent.
- SYSCOR's complex scoring system makes it difficult to remove or add variables, or to change the number of categories, or the assignment of a variable to a class, if scientific understanding about its importance changes. It is, however, easy to change the classification categories.

Pest management

Losses of agricultural production because of pests can jeopardise farm economic viability. Pesticides are generally used when the financial benefit, measured by the value of increased yield or crop quality, exceeds the cost of applying the pesticide. Pest management decisions mainly involve applying the mix of pesticides more efficiently and choosing between biological pest control methods and pesticides. Where pesticides are used, the objective of reducing the cost of pesticide use is achieved through decisions which involve selecting the most appropriate pesticides, the timing of the application and the application method.

Insect monitoring is widely used to determine the timing and frequency of insecticide application, and the same method can be applied for fungal diseases. Fungicides are also often applied to seed as an insurance against subsequent cool, wet conditions that would encourage fungal disease of the seedlings. The decision to use these fungicides is often made by the seed producer, and it can be difficult to obtain untreated seeds.

Monocultures in arable production often increase pest problems and the risk of strains of insects and weeds developing resistance to pesticides. Inclusion of forage crops with grain or horticultural crops, in regular crop rotation, is likely to reduce the need for pest control. Allelopathic crops and residues release natural compounds that discourage certain weeds. Some insect pests are repelled by certain plants and materials made from naturally occurring hormones.

Non-chemical pest control methods

Indicator definition and method of calculation
The indicator shows the area that has not been treated with chemical pesticides, and is calculated as the crop area that is not treated with chemical pesticides divided by the total cultivated agricultural area. The cultivated agricultural area includes the total arable and permanent crop land, assuming that pesticides are not used on temporary or permanent pasture. Non-chemical pest control methods include, for example,

tillage (e.g. ploughdown of allelopathic residues, that is, plants whose roots and residues can suppress the growth of many other plants, including weeds), crop rotation, biological control (e.g. parasitic organisms for control of insect pests), pheromones and hand weeding.

Recent trends

Chemical pesticides are not used in organic farming; hence Fig. 8.4, showing trends in the share of agricultural land under organic farming, can also be considered to reflect trends in the area where only non-chemical pest control methods are used. Organic farming systems also include many other requirements and, consequently, the area where chemical pesticides are not used often exceeds the area under organic farming. Examples of such countries include Germany, Spain and the United Kingdom, where significantly more farmers are now using non-chemical pest control methods than in the 1980s.

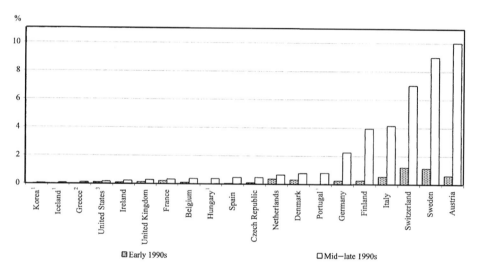

Fig. 8.4   Share of the total agricultural area under organic farming: Early 1990s and mid–late 1990s (OECD, 2001).

Notes:
1. Data for the early 1990s are not available.
2. Percentage for the early 1990s equal 0.003%.
3. Data for the United States are taken from Welsh (1999).

Over a third of Canadian farms and field crop area do not apply commercial pesticides (Table 8.2). In addition to the use of non-chemical pest control methods, Canada has developed other indicators for pesticide management (timing of herbicide applications, timing of insecticide and fungicide applications and sprayer calibration) (McRae *et al.* 2000). The indicators suggest that herbicide application was triggered by the level of economic injury to the crop on about 20% of treated crop land. Also farmers were more likely to apply herbicides at a certain stage of crop growth or to use the first sign of pests to time pesticide applications. Moreover, nearly 70% of farmers calibrated sprayers only at the beginning of the crop season (McRae *et al.* 2000).

Table 8.2 Pest control methods used by farmers excluding the use of chemical pesticides, Canada, 1995 (McRae *et al.* 2000).

Pest Control Method	Number of farms	% of farm numbers[1]	% of field crop area treated[1]
Tillage	53.805	26	28
Crop rotation	99.970	49	56
Biological control	4.570	2	2
Pheromones	495	< 1	< 1
Hand weeding	14.900	7	4
Other	2.605	1	1
No non-chemical method	80.510	39	34

[1]Percentages may exceed 100% where more than one practice is used on the same crop area.

The pest management practices included in the indicator are assumed to pose fewer risks to human health and the environment than 'conventional' pesticide application methods and they can potentially be applied to manage pest pressures without affecting farm profitability. The definitions of practices need to be harmonised to improve international comparability and the data availability needs to be improved.

In general it can be assumed that an increase in agricultural area under non-chemical pest control methods is good for the environment. However, some caution is required with such an interpretation, as it will be necessary to link these farm management practices to actual environmental outcomes, or outcomes measured through other indicators such as soil and water quality, biodiversity and wildlife habitats.

Integrated pest management (IPM)

Indicator definition and method of calculation
This indicator measures the area under IPM[10] divided by the total cultivated agricultural area. The cultivated agricultural area includes the total arable and permanent crop land, assuming that pesticides are not used on temporary or permanent pasture. IPM is a knowledge-intensive and farmer-based management approach that encourages natural control of pest populations by anticipating pest problems and preventing pests from reaching economically damaging levels. Activities under IPM include, for example, the enhancement of natural enemies, planting pest-resistant crops, adapting cultural management, and 'judicious' use of pesticides.

Recent trends
From the limited information that exists, it appears that significantly more farmers are now using integrated pest management (IPM) than in the 1980s. In the United States, IPM was applied on over 50% of the fruit, vegetable and major field crop (maize and soybeans) area in the early 1990s (Vandeman *et al.* 1994).[11] Scouting for insects and diseases is already used on 75% of fruit crops and nearly 75% of vegetable crops (OECD 1997). A number of these farmers also used pest-resistant crops, cultural management and other non-chemical techniques. The United Kingdom does not record the IPM area separately, but a survey

prepared by the United Kingdom Department of the Environment in 1997 estimated that 50% of farmers use IPM techniques on their farms (MAFF 2000).

Indicator interpretation and limitations
Indicators of IPM, as for non-chemical pest control methods, are assumed to pose fewer risks to human health and the environment than 'conventional' pesticide application methods and can potentially be applied to manage pest pressures without affecting farm profitability. The definitions of practices need to be harmonised to improve international comparability, and the data availability on areas where both chemical and non-chemical methods are used in parallel, including IPM, needs to be improved.

The cultivated area under IPM is an indicator of comprehensive pest management, reduced pesticide risk, and optimal timing of pesticide use (as measured by the number or area of farms and/or crops where IPM is used). It addresses all pests and pest control methods, and it attempts to optimise the use of pesticides, not to replace them. It may be the best indicator of farm pest management efficiency, but it probably has a lower sensitivity to environmental concerns than the indicator on the use of non-chemical pest control methods.

It is necessary to distinguish between certain herbicides and other pesticides. This is partly because the mode of action and potential toxicity to non-target organisms of herbicide use is generally less environmentally hazardous compared to the consequences of using other pesticides. It is also because herbicides are frequently used to reduce tillage, which has considerable environmental benefits. Herbicide materials can be divided into those that are used in forage or close-grown crops, where there is no benefit from reduced tillage, and those used primarily in wide-row crops, and in reduced or no-tillage systems, as an alternative to tillage.

Future research challenges

A key aspect to future research in developing pesticide indicators is to improve the collection, coverage and quality of pesticide use and/or sales data, expressed in terms of the quantity of active ingredients. This work might also include collecting information on pesticide use per crop per hectare. Incomplete data on pesticide use can be a significant obstacle to development of meaningful risk indicators.[12]

The initial focus of the OECD pesticide risk indicators project is on methods for calculating indicators of aquatic risks. Indicators for human and terrestrial risks will follow. A recent OECD survey that identified and described existing pesticide risk indicators developed by OECD countries, and work already completed by several countries, will provide a starting point for this work.[13] The basic approach for all risk areas will be to combine information on pesticide hazard and exposure (i.e. risks) with information on pesticide use and/or sales. The project is not seeking to combine the indicators of human health and environmental risks into one 'general' indicator of pesticide risk trends, as OECD countries consider such an approach scientifically invalid.

As work on pesticide risk indicators develops, however, it will be important to strike the right balance. On the one hand there is a need to develop a simple risk assessment system drawing on readily available data and research, which can be improved over time. On the other hand, developing a more comprehensive system of risk indicators, which may have greater scientific accuracy, can be difficult to manage in terms of its complexity and data requirements, and may not be easily understood by policymakers and other stakeholders. Moreover, pesticide risk indicators need to be related to other agri-environmental indicators rather than pesticides used alone, especially those covering farm pest management, soil and water quality, and biodiversity (see Fig. 8.1).

Data availability is the main barrier to wider use of pest management indicators, as many OECD countries do not have reliable information on the extent to which these practices are used. Environmental conditions and farming systems vary within and across OECD countries and, consequently, best farm management practices vary from one region to another. For example, there is no need to change pest control practices if pesticide use is already at a low level for climatic or other reasons. Thus, identifying and developing a standard set of indicators on pest management practices across the OECD is not straightforward.

It may also be useful in the future to supplement physical indicators of changes in pesticide use and risks and pest management practices with economic indicators. This might be achieved by exploring the possibility of developing a 'cost:benefit' approach that analyses the relationship between the environmental and health costs associated with pesticide use, and the benefits derived from pesticides in terms of improvements in agricultural productivity (Pearce and Tinch 1998). At present the scale of the costs relative to the benefits of pesticides is uncertain, and it is this relative economic assessment which is needed better to guide policymakers and inform the public.

Acknowledgements

The authors would like to thank colleagues at the OECD Secretariat for help in preparing this chapter, especially Laetitia Reille, Wilfrid Legg, Outi Honkatukia, Jeanne Richards, Françoise Bénicourt and Theresa Poincet. Any remaining errors in the chapter are the responsibility of the authors, and the views expressed do not necessarily reflect those of the OECD or its member countries. The text of this chapter is based on OECD (2001) (see the OECD website: www.oecd.org/agr/env/indicators.htm. See also OECD (1999a) and OECD (1999b)).

Notes

1 The thirty OECD member countries include: Australia, Austria, Belgium, Canada, Czech Republic, Denmark, Finland, France, Germany, Greece, Hungary, Iceland, Ireland, Italy, Japan, Korea, Luxembourg, Mexico, Netherlands, New Zealand, Norway, Poland, Portugal, Slovak Republic, Spain, Sweden, Switzerland, Turkey, United Kingdom, United States.

2 Mexico removed its use of pesticide subsidies from 1998, while the Czech Republic, Hungary and Poland also used pesticide subsidies prior to 1990.

3 For a review of OECD pesticide policies and the environment, see OECD (1997).

4 The UNEP has published a report on phasing out ozone depleting methyl bromide, which includes an extensive international database (UNEP 2000).

5 For related studies that have examined the links and related indicators covering pesticide use and risks and other agri-environmental areas, see, for example, Commonwealth of Australia (1998); ECNC (2000); European Commission (1999); MAFF (2000); and USDA (1997).

6 In Germany the use of plant protection products relating to agricultural areas would be reduced by approximately 30% over the period concerned in Fig. 8.2, but data for the former East Germany are not available.

7 In Australia and New Zealand, where pesticide use data time series are incomplete, pesticide use indicators are now being developed; see, for example, Hamblin (1998), for Australia; and Holland & Rahman (1999), for New Zealand.

8 Further information on the OECD's work on pesticide risk indicators is available on the OECD website at: http://www.oecd.org/ehs/ [Pesticide Programme > Pesticide Risk Reduction]. For a review of other work on pesticide risk indicators, see, for example, CAE (1999); Falconer (1998); and Oskam & Vijftigschild (1999).

9 In the project, SYSCOR was not designed to calculate long-term risk indicator, but could be modified to do so.

10 The OECD also held a Workshop in 1999 in Switzerland on Integrated Pest Management (for further details, see the OECD website at: http://www.oecd.org/ehs/ [Pesticide Programme > Pesticide Risk Reduction]).

11 Farmers were considered to be using IPM 'if, before making pesticide application decisions, they monitored pest populations (scouting) in order to determine when a pest population had reached an economically damaging threshold'.

12 OECD in cooperation with EUROSTAT is beginning a process to improve quality and coverage of pesticide use data.

13 For details of this survey and the future OECD Programme of Work on pesticide risk indicators, see the OECD website at: http://www.oecd.org/ehs/ [Pesticide Programme > Pesticide Risk Reduction].

References

CAE [Centre for Agriculture and the Environment] (1999) *Comparing environmental risk indicators for pesticides.* Results of the European CAPER Project, CLM Report No.426. CAE, Utrecht [summary at: http://www.clm.nl/index_uk2.html].

Commonwealth of Australia (1998) *Sustainable agriculture: Assessing Australia's recent performance.* Report to the SCARM of the National Collaborative Project on Indicators for Sustainable Agriculture, SCARM Technical Report No.70. CSIRO Publishing, Victoria.

Department of the Environment (1996) *Indicators of Sustainable Development for the United Kingdom.* DoE, London.

ECNC [European Centre for Nature Conservation] (2000) *Agri-environmental Indicators for Sustainable Agriculture in Europe.* ECNC, Tilburg.

EEA [European Environment Agency] (1995) *Europe's Environment: The Dobris Assessment.* European Environmental Agency, Office for Official Publications of the European Communities, Luxembourg.

EEA [European Environment Agency] (1998) *Europe's Environment: The Second Assessment.* Office for Official Publications of the European Communities, Luxembourg. [Available at: http://themes.eea.eu.int/ [> all available reports]].

European Commission (1999) *Agriculture, Environment, Rural Development: Facts and Figures. A Challenge for Agriculture.* Office for Official Publications of the European Communities, Luxembourg. [Available at: http://europa.eu.int/comm/dg06/envir/report/en/index.htm]

EUROSTAT (1999) *Towards environmental pressure indicators for the EU.* Environment and Energy Paper Theme 8, Statistical Office of the European Communities, Luxembourg. [Background documentation at: http://e-m-a-i-l.nu/tepi/ and http://esl.jrc.it/envind/]

Falconer, C. (1998) *Classification of pesticides according to environmental impact.* Final Report to the RSPB. Centre for Rural Economy, Department of Agricultural Economics and Food Marketing, University of Newcastle-upon-Tyne.

Fernandez-Cornejo, J. & Jans, S. (1999) *Pest management in US agriculture.* Agricultural Handbook No. 717, Resource Economics Division, Economic Research Service, US Department of Agriculture, Washington, DC. [Available at: http://www.ers.usda.gov/ [Publications > Inputs and Technology > Fertilizer and Pesticides]]

Gebbie, D. (1998) Chemical issues in international trade. In: *Proceedings of the Australian National Pesticide Risk Reduction Workshop* (eds P. Rowland & D. Bradford), pp. 11–19. Bureau of Rural Sciences, Agriculture, Fisheries and Forestry, Canberra.

Hamblin, A. (1998) *Environmental Indicators for National State of the Environment Reporting: the Land.* State of the Environment (Environmental Indicator Reports) Department of the Environment, Canberra. [Available at: http://www.environment.gov.au/soe/ [Environmental Indicators > Land under 'Environmental Indicator Reports']]

Holland, P. & Rahman, A. (1999) *Review of trends in agricultural pesticide use in New Zealand.* MAF Policy Technical Paper 99/11, Ministry of Agriculture and Forestry, Wellington. [Available at: http://203.97.170.4/MAFnet/index.htm [Site A-Z > T > Technical Papers]].

IFEN [Institut français de l'environnement] (1998) *Pesticides in Water.* Etudes et Travaux no. 19, IFEN, Orléans, France. [Available at: http://www.ifen.fr/pestic/pestic.htm]

MAFF [Ministry of Agriculture, Fisheries and Food] (2000) *Towards Sustainable Agriculture: A Pilot Set of Indicators.* MAFF, London. [Available at: http://www.maff.gov.uk/farm/sustain.htm]

Mano, S. & Andreae, M.O. (1994) Emission of methyl bromide from biomass burning. *Science,* 263 (4 March), 255–6.

McRae, T., Smith, C.A.S. & Gregorich, L.J. (2000) *Environmental Sustainability of Canadian Agriculture: Report of the Agri-environmental Indicator Project.* Agriculture and Agri-Food Canada, Ottawa. [http://aceis.agr.ca/policy/environment/sustainability/performance/indicators/aei.html].

Oberthur, S. (1997) *Production and Consumption of Ozone Depleting Substances 1986–1995.* Deutsche Gesellschaft fur Technische Zusammenarbeit, Berlin.

OECD [Organisation for Economic Co-operation and Development] (1997) *Agricultural Policy, Pesticide Policy and the Environment.* OECD, Paris.

OECD [Organisation for Economic Co-operation and Development] (1998) *The Environmental Effects of Reforming Agricultural Policies.* OECD, Paris.

OECD [Organisation for Economic Co-operation and Development] (1999a) *Environmental*

Indicators for Agriculture Volume 1: Concepts and Framework. (Reprinted from the first edition of 1997) OECD, Paris.

OECD [Organisation for Economic Co-operation and Development] (1999b) *Environmental Indicators for Agriculture Volume 2: Issues and Design. The York Workshop.* OECD, Paris.

OECD [Organisation for Economic Co-operation and Development] (1999c) *OECD Survey on the Collection and Use of Agricultural Pesticide Sales Data: Survey Results.* OECD, Paris. [http://www.oecd.org/ehs/ [Publication > Pesticides]]

OECD [Organisation for Economic Co-operation and Development] (1999d) *OECD Environmental Data Compendium.* OECD, Paris.

OECD [Organisation for Economic Co-operation and Development] (2001) *Environmental Indicators for Agriculture Volume 3: Methods and Results.* OECD, Paris.

Oskam, A. & Vijftigschild, R. (1999) Towards Environmental Pressure Indicators for Pesticide Impacts. In: *Environmental indicators and agricultural policy*, (eds F. Brouwer & B. Crabtree), pp. 157–176. CAB International, Wallingford.

Pearce, D. & Tinch, R. (1998) The True Price of Pesticides. In: *Bugs in the system. Redesigning the pesticide industry for sustainable agriculture* (eds W. Vorley & D. Keeney), pp.50–93. Earthscan, London.

Rayment, M., Bartram, H. & Curtois, J. (1998) *Pesticide Taxes: A Discussion Paper*. Royal Society for the Protection of Birds, Sandy.

UNEP [United Nations Environment Program] (2000) *Methyl bromide Phase-out Strategies: A Global Compilation of Laws and Regulations.* UNEP. [http://www.uneptie.org/ ozonaction.html [Sector-specific Information > Methyl Bromide > Policy Information]]

USDA [United States Department of Agriculture] (1997) *Agricultural Resources and Environmental Indicators, 1996–97.* Agricultural Handbook No. 712. Economic Research Service, Natural Resources and Environment Division, Department of Agriculture, Washington, DC. [[http://www.ers.usda.gov/ [Briefing Rooms > Agricultural Resources and Environmental Indicators]]

USGS [United States Geological Survey] (1999) *The quality of our nation's waters: Nutrients and pesticides.* US Geological Survey Circular 1225. United States Geological Survey, Washington, DC. [http://water.usgs.gov/pubs/circ/circ1225/]

Vandeman, A., Fernandez-Cornejo, J., Jans, S. & Lin, B. (1994) *Adoption of integrated pest management in US agriculture.* AIB No. 707, September 1994. Economic Research Service, Department of Agriculture, Washington, DC.

Welsh, R. (1999) *The Economics of Organic Grain and Soyabean Production in the Midwestern United States.* Policy Studies Report No. 13. Henry A. Wallace Institute for Alternative Agriculture, Maryland. [http://www.hawiaa.org/hawiaa.htm]

Chapter 9

Integrated Assessment of Pesticides

Methods for Predicting and Detecting Environmental Risks in a Safety Net

Harrie A.J. Govers, Pim de Voogt, Pim Leonards, André van Roon and Onno Kwast

Introduction

In this chapter we discuss a proposed pesticide safety net which covers all stages of the life cycle of a pesticide. Emphasis is on natural science elements relevant to registration policies (in a broad sense) via *ex ante* assessment during design and (temporary) admission and via *ex post* assessment of environmental impacts. Unfortunately, potential impacts of new active ingredients never can be fully predicted and detection in the environment remains indispensable. Therefore, the proposed net combines prediction and detection. A chemical's reactivity is a key property in this respect. It plays a role in all stages of the life cycle, leading to impurities in agrochemical products and metabolites in the environment. In this chapter we review the availability of methods of prediction and detection of environmental occurrence and effects of pesticides, their impurities and metabolites.

We also describe and analyse two examples of integration and registration policies. The first one treats a chlorinated pesticide currently in use, 4-chloro-2-methylphenoxy acetic acid (MCPA). Here, we examine an early stage prediction method for the formation of unwanted side products (impurities) during its production. The second example deals with monoterpenoids, potential substitutes for pesticides currently in use and sustainably produced. In this case, especially, we discuss the prediction of biodegradation.

This chapter focuses on the natural scientific improvements in and alternatives to the integrated registration (temporary admission) and assessment of (organic) pesticides. As such it can be considered as a contribution to alternatives to current registration and decision procedures (*cf.* Vogelezang-Stoute in this volume). In addition, this chapter starts with the assumption that the production, use and environmental impact of pesticides have been causing many problems that need to be solved (Struik & Kropff in this volume).

Organic pesticides, especially the chlorinated ones, have been a subject of scientific and political debate since the publication of *Silent Spring* in 1962 by Rachel Carson. In the past and for reasons of persistency and toxicity, chlorinated organic chemicals such as p,p'-DDT were developed in order to be applied as pesticides. Many of these, though banned in highly industrialised countries, are in use in developing countries, and chlorinated pesticides are still used in industrialised countries. In The Netherlands, for example, about 300 active compounds (active ingredients, ais) are currently admitted (CTB 1998, personal communication; den Hond in this volume). Among these are chlorinated compounds such as mecoprop-P en MCPA. In the societal debate, chlorinated pesticides form an essential part of the total group of chlorinated compounds (CML & TNO 1995; Tukker 1998). Such debate leads to positions aiming either at the complete ban of chlorinated organics or at a continuous risk evaluation of each new or old compound separately, eventually followed by a ban. Connected to this, the concept of 'sustainable development' (SD) has been defined by the international debate (WCED 1987) as: 'Development that meets the needs of the present without compromising the ability of future generations to meet their own needs'. An important implication of the concept in chemical and pesticide industry, in 'green' or 'soft' chemistry, is the saving of raw material and energy and the reduction of environmental pollution by the use of renewable resources such as biomass-based chemicals (DTO 1997; van Roon *et al.* 2001). However, the potential benefits of these compounds with respect to the reduction of environmental risks should be evaluated equally thoroughly as those of the traditional compounds.

The key property of any chemical is its reactivity, that is, its formation and transformation, under the pertinent conditions met during its entire life cycle. Under the artificial conditions of industrial production, reactivity of basic chemicals leads to the formation of the pesticides sought for, but often accompanied by, impurities such as products of unwanted side or subsequent reactions and traces of unreacted basic chemicals or solvents. In the subsequent formulation of the ai, its transport to agricultural users and its open applications further reactions may occur. In the environmental compartments of soil, sediment, water and air a lack of reactivity (persistency) may lead to dispersion of the chemical over large distances and an accumulation in organisms (van Dijk *et al.* 1999). Here, environmental conditions such as temperature, aerobicity, presence of water and acids, sunlight and sorbing media determine the transformation of compounds in addition to their intrinsic properties. Moreover, when taken up by biota, the chemical may undergo biochemical reactions catalysed by the enzyme systems of the pertinent organisms. The result may be beneficial, as in complete biodegradation by bacteria or fungi, or harmful, when toxic metabolites are produced in higher organisms.

From the viewpoint of environmental policy the behaviour of a new pesticide should be predictable with respect to the concentration of active ingredients, impurities and metabolites built up in time in the various environmental compartments. To this end fate models can be used (van de Meent 1993) in which reliable input data on reactivity, degradation rate constants, are of eminent importance. In addition, it should be known what doses in these compartments are toxic to specified organisms and these should be determined by laboratory experiments or prediction methods. Due to the complexity of environmental conditions and to the diversity of organisms, these predictions will never be complete, despite progress made in predictive environmental sciences. As a consequence, chemical and biological detection methods of original pesticides, impurities and metabolites remain indispensable and should be developed and applied, leading to a timely re-evaluation of the admission or registration of a pesticide. Of course, detection methods are also important when the pesticides are used without permission and no check for the availability of low limit detection methods could take place during registration.

In the second section, we outline a pesticide safety net which integrates prediction and detection methods, with emphasis on natural science elements relevant to policy-making (registration policy). The concept of the safety net was developed in a recently finalised multidisciplinary project, PROMPT, on the prevention of risks from organic micropollutants (Govers 1997). The safety net will be compared to the existing registration procedures and policy decisions in both governmental and corporate institutions. In addition, this section will briefly discuss the shortcomings of current procedures and possible improvements of the net proposed. In the third and fourth sections we review methods of prediction and detection of environmental occurrence and effects of pesticides, their impurities and metabolites. Two examples are given and analysed with respect to the registration procedures or policy decisions of the safety net. The fifth section treats a chlorinated pesticide currently in use, 4-chloro–2-methylphenoxy acetic acid (MCPA). Here, we emphasise an early stage prediction method for the formation of unwanted side products (impurities) during its production. The sixth section deals with potential pesticide substitutes currently in use and sustainably produced: the monoterpenoids. Here, we deal especially with the prediction of biodegradation. In the final section, with discussion, conclusions are drawn on the integration of methods and the registration and decision policies of the safety net.

Outline of an integrated safety net

The preventive safety net proposed is defined as 'a set of scientific techniques and environmental policy measures plus practical procedures of societal actors into which these techniques have been incorporated' (Govers 1997). Its objective is to prevent environmental risks as early as possible at all crucial stages of the entire societal route of a compound and its appearance in the environment. Five stages are distinguished:

(1)   research and development or design;
(2)   marketing and production;
(3)   formulation, packaging, storage and transport;
(4)   *in situ* agricultural use and storage;
(5)   environmental behaviour and early detection.

As depicted in Fig. 9.1, registration and assessment policy (RP) measures include: RP0 (during industrial design in the first stage), RP1 (temporary admission of bringing on the market at the end of the second stage) and RP2 (re-evaluation of RP1 registration and/or decision to develop new compounds – RP0 – based on detection in the environment). Societal actors directly connected to each of the stages are, among others: researchers and research managers of a company (stage 1), production workers, managers and marketing departments of companies and governmental registration and working place hygiene control committees (stage 2), selling departments, formulation companies and transportation firms (stage 3), agricultural firms and employees, governmental working place hygiene controllers (stage 4), and consumers, including local environmental pressure groups (stage 5). In addition indirect actors are involved, such as general administrational bodies in the companies, general governmental authorities and consumer plus environmental organisations.

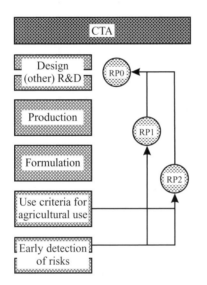

- Constructive technology assessment (CTA)
- New RP0 'Registration' in addition to RP1 and RP2
- Emphasis on design, use criteria and early detection
- *Ex ante* and *ex post* assessments

Fig. 9.1   PROMPT pesticide safety net.

Elements of registration or admission decisions to be taken by industrial managers or the governmental authorities, or both, already exist in industrialised countries; *ex ante* decision in RP1 is especially well developed. Admission or permission to bring on the market is temporary, lasting for a period of, for example, ten years. Decisions on the further development of active ingredients are taken by industrial managers on the basis of a comparison to (governmental) registration criteria, and governments or governmental agencies who take registration decisions (*cf.* Irwin & Rothstein in this volume). It is up to the companies either to fulfil the additional data requirements or to withdraw the registration request. Large industrial corporations increasingly apply formalised internal *ex ante* design procedures for pesticides and other chemicals (Plummer 1990; de Vito & Garrett 1996; AKZO 1997, personal communication). Somewhere in the procedure a statement of 'no objection against further development' is included (SNOB, RP0) (see Fig. 9.2). The column of boxes in Fig. 9.2 denotes the general procedure for the design of an active ingredient compound. An important element is the search for compounds or chemical structures with potentially high biological activity, which could be the lead ('LEAD') to further optimisation of wanted compound properties. Connected to this is a box including environmentally relevant data to be known for risk assessment. On the upper right side of the figure the

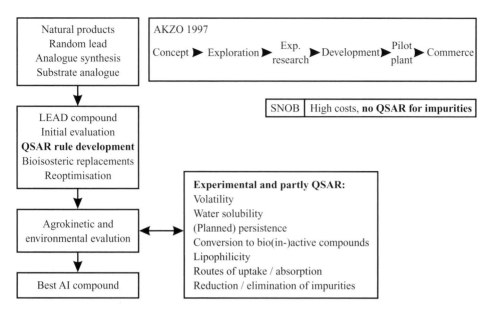

Fig. 9.2 Corporate design procedures (Plummer 1990; de Vito & Garrett 1996; AKZO 1997, pers. comm.).

general AKZO procedure is given for the development of chemicals. Each of the boxes includes statements with respect to the use of QSAR (Quantitative Structure–Activity Relationship) prediction techniques.

However, these procedures are not yet generally used. Moreover, they are limited in their ability to predict impurity formation and planned persistency, as will be considered later in this chapter. A final drawback of current procedures is the weakness of the RP2 *ex post* evaluation based on detection and monitoring of compounds, impurities and metabolites in the environment, which will also be elaborated upon later. Renewed admission considerations in RP1 leading, for example, to the ban of chemicals are scarce, time consuming and seldom based on regular data monitoring as a main feature. Moreover, many compounds, impurities and, especially, metabolites are not being monitored. A further criticism considers the degree of 'constructive technology assessment' (CTA) as currently applied. In no way is a general consideration of biological, mechanical and other non-chemical means of pest control included in the current registration procedures (Adams 1995; Schot 1996). CTA would also include the hypothesis that societal problems with technology can be dealt with by broadening the design process of technology. This may be done through the inclusion of societal actors and their views, perspectives, hopes and wishes into the design process. Broadening has been limited by agrochemical companies to the technical aspects of integrated pest management (IPM) or integrated crop management (ICM). CTA will not be considered in this chapter. However, the introduction of CTA would also increase the pressure on governmental and corporate institutions to improve access to scientific assessment data of pesticides for both the general public and non-corporate research institutions. The limited accessibility of this type of data did hamper the completeness of this chapter.

Prediction methods

The procedure for the assessment of risks of new chemical compounds includes as a first step the prediction of environmental concentrations by fate models such as the Simplebox model (van de Meent 1993) or the fugacity model (Mackay & Paterson 1990). These models are only partially validated by comparison of predicted data with field data. In particular, the predictions of concentrations in soil and sediments and in certain water compartments easily deviates by a factor of 100 or more from field data for compounds currently in use (Cowan *et al.* 1995). A main cause of inaccuracy is the incomplete inclusion of the environmental system and its properties (e.g. artificial boundaries of the system, heterogeneity of compartments being considered as homogeneous, incomplete inclusion of biota, lack of data on bulk flows of air, water and sediments). Thus the system is mostly of a very hypothetical character. Other causes of inaccuracy are connected to the compound (pesticide) itself: lack of data on the amount of the chemical emitted to the environment and lack of (laboratory) data on process parameters such as volatility, desorption constants, dissolved organic matter–water partition constants and, as mentioned before, reliable degradation constants. In most of the cases, when one or more experimental data on compound properties are unavailable, prediction of properties is required (see below).

The second step is the comparison of these predicted concentrations (PEC) with predicted (or in the laboratory, measured) 'no observed toxic effect' levels or other (eco-)toxicological standards (van Leeuwen & Hermens 1995). Toxicological input data may be unavailable or inaccurate. These two steps are currently included in the European EUSES model (ECB 1997) for the assessment of chemicals in general.

For pesticides and groundwater contamination in The Netherlands the models of PESTLA (Boesten & van der Linden 1991) and its successor PEARL (Leistra *et al.* 2000) are used. For surface water contamination SLOOT.BOX (Linders *et al.* 1990) and its successor TOXSWA (Adriaanse 1996) are available. The model of PESTRAS (Tiktak *et al.* 1994) is used rarely by CTB in cases where the normal model to estimate concentrations in groundwater was shown to give incorrect answers because of too high vapour pressure of the active ingredient under investigation.

Table 9.1 Experimental environmental chemistry data, required per environmental compartment, for the assessment of risks and to be obtained by accepted methods (Lynch 1995).*

Soil	Water	Air
aerobic degradation (DT_{x0})	aerobic degradation (route)	photolysis (k)
anaerobic degradation (DT_{x0})	anaerobic degradation —	
photolysis (DT_{x0})	hydrolysis (k)	
field dissipation (DT_{x0})	photolysis (k)	
field residue study	fish bioaccumulation (BCF)	
ad/de-sorption (K_d, K', K_{oc})		
leaching studies		
aged residue leaching		
field leaching		
volatility		

*Also toxic effect parameters are required and available.

Experimental data for the assessment of pesticides obtained by accepted methods are summarised in Table 9.1.

The (limited) availability of prediction methods for pesticide properties required for fate models is summarised in Table 9.2 together with some limitations on the accuracy of that property which can be predicted most accurately of all properties: the partition constant of a compound over n-octanol and water (Kow) (Fig. 9.3). As shown in Fig. 9.3, the error tends to be within one log-unit for the latter property. However, this error may easily increase to a factor of 1000 for properties such as degradation rate constants.

Table 9.2   Availability of QSAR methods for the prediction of environmental chemistry data.

QSARs are more (+), less (o) or not (-) available for the fate properties:

Vapor pressure	(o)
Aqueous solubility	(o)
Kow	(+)
Koc	(o)
BCF in fish	(o)

Rate constant soil (s) / water (w) / air (a)	
anaerobic degradation	s (o), w (-)
aerobic degradation	s (o), w (o)
photolysis	s (-), w (o), a (+)
hydrolysis	w (+)

Prediction methods for properties, also for toxic effect levels, mostly apply the principle of Quantitative Structure–Activity Relationships (QSARs). In the latter a mathematical relationship (equation) between a certain compound property to be predicted (the dependent variable) and one or more known compound properties (the structural or physicochemical independent variables) is established for a training set

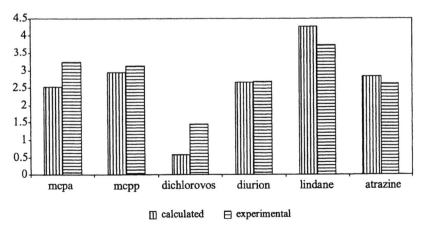

□ calculated   ⊟ experimental

Fig. 9.3   LogKow of six chlorinated pesticides, calculated by Rekker's method.

of compounds, the dependent variable of which is known by experiment. Once this relationship has been established the dependent variable of compounds with lacking experimental data can be predicted by insertion of the known independent variables of the compound into the QSAR equation. During the last decades QSAR has become a specific field of study and a method applied in many studies reported in text-books and handbooks (e.g. Karcher & Devillers 1990; Mackay & Boethling 2000) and journals such as *SAR and QSAR in Environmental Research*. QSAR prediction methods tend to be inexpensive and fast compared to the experimental determination of properties.

Detection methods

Generally speaking two quite different types of field detection methods for organic compounds are available and under continuous development: chemical analysis and biomarking plus other biological methods. As biomarking is the newer one we will pay it special attention. Detection will be considered to include both the identification and the quantification of a compound.

In *chemical analysis* (Harrison *et al.* 1993) one or more samples of material (the matrix) of soil, sediment, water, air or biotic materials are taken, conserved, stored, cleaned up and prepared, and injected into an equipment for analysis, that is, for identification and quantification. Then data collecting and interpretation take place and conclusions on type and concentrations of compounds are drawn. The number and types of sampling are overwhelmingly large and very specific for the types of compound under investigation, as are the number and types of equipment for analysis. Modern procedures such as those based on gas chromatography combined with mass spectrometry (GC-MS) enable the identification and quantification of many compounds of a similar type simultaneously, including new and unexpected ones. They are very compound-specific and have low detection limits. Modern procedures also include the combination of methods such as gel permeation chromatography followed by gas chromatography with electron capture and mass spectrometric detection (Gelsomino *et al.* 1997). In addition, methods are available which characterise complex mixtures by sum parameters such as extractable organohalogens (EOX). Here no information is obtained on separate compounds, although this would be required from a toxicological point of view and in order to trace the source of the pollutant. Chemical analysis itself does not provide information on biological effects. To this end toxic levels have to be determined separately. Chemical analysis of environmental pollutants is a very well-established and developed branch of chemistry, often applied routinely under rigorous quality control, including the statistics of sampling and analysis. New developments focus on: new compounds (e.g. metabolites), lower detection limits, higher accuracy, hyphenation (coupling of procedures and/or instruments), faster procedures (automation), automation of data interpretation, use of databases, miniaturisation of equipment, high user friendliness and lower costs. Hyphenation tends to improve both the specificity and the scope of compounds to be detected in a single run. Other detection methods, such as the biomarkers treated below, could supplement or substitute chemical analysis, especially when they meet these criteria in a better way.

Biomarkers require a similar way of sampling. However, identification and quantification is quite different and based on the biological response of a whole organism or parts of it. This response is thought to give information on the type, amount and biological activity of compounds or susceptibility of organisms under field conditions. In current definitions the biomarker is confined to the biological level of biochemical, physiological, histological and morphological processes (van Gestel & van Brummelen 1996). Incorporation of the biomarker concept in ecotoxicology calls for a redefinition of terms. On the level of an individual organism, toxic endpoints such as mortality and effects on reproduction in the field are studied in bioassays. Biomarkers show a gradient in suitability from exposure detection to detection of effects as shown in Table 9.3, where true biomarking is confined to the levels (a)–(f). From the view point of the safety net, biomarking could be evaluated according to the following criteria (Oikari *et al.* 1993):

(1) high sensitivity: evocable and measurable following a short exposure to low concentrations of the contaminant;
(2) a clear relationship to the concentration of the contaminant; a well defined dose–response curve;
(3) a high specificity: non-responsive to confounding factors arising from natural environmental events and normal physiological changes in the sentinel species, which also includes a low variability between individuals both within a single species and across species;
(4) based on a validated, published method that can produce similar results at any time in any qualified laboratory. The method should be validated in the field;
(5) integrate the response from all possible exposure routes;
(6) based on a well-characterised biochemical or physiological mechanism that would provide the understanding needed to interpret the biological and ecological consequences of the response. The measured change in the organism should be one that, under continued contaminant exposure, would result in a population- or community-level effect. Ideally, however, the sensitivity of the biomarker response should be large enough to measure this change before irreversible ecological consequences occur.

In addition to the criteria for improvement of chemical analysis, as given above, biomarkers ideally have the potential for simultaneous detection of contaminants and their effects. In this way laboratory experiments for the determination of toxic levels, required in the case of chemical analysis, could perhaps be avoided.

A recent review (de Knecht & van Brummelen 1998) demonstrates that biomarkers currently fulfil these criteria only partially. Large problems are met with respect to clear dose–response curves, combination of high sensitivity with high specificity, the occurrence of confounding factors and the lack of knowledge of mechanisms for organisms to be selected. Positive exceptions are mainly confined to biomarkers for aquatic ecosystems.

Table 9.3   Gradient of biomarking from exposure to effect detection (after DiGiulio *et al.* 1993).

(a) Biological changes of no known adverse effect associated with exposure

(b) Biological effects that could comprise the organism (EROD induction)

(c) Biochemical, cellular, or physiological changes clearly indicative of an adverse effect but of uncertain eventual consequence (protein or DNA adduct)

(d) Biochemical, cellular, or physiological changes clearly indicative of a toxic effect of known consequence based on mechanism understanding (inhibition of AchE)

(e) Structural tissue/organ disorder in individuals (necrosis/tumours)

(f) Clear detriment to the organism (liver and gonadal degeneration, death)

(g) Assessment endpoints such as clear detriment to population.

Similar to chemical analysis, monitoring programmes are recommended to make use of a so-called tiered approach starting with relatively inexpensive and rapid non-specific general biomarkers (responding to a wide variety of chemicals). Subsequently, if any of these indicate that a hazard exists, more specific biochemical biomarkers can be used to diagnose the cause of the environmental stress. When high ecological relevance is required, high level organisms should be used, which, however, show low sensitivity in contradiction to the aim of an early warning safety net. Thus, currently the first tier could include the use of biomarkers of the lowest level (DNA integrity, lyosomal stability, MFO induction or stress proteins) with high sensitivity, low costs, high speed and ease, but low specificity and ecological relevance. Suitable biomarkers of this type to be used among others for organic pesticides (OP) are summarised in Table 9.4 (de Knecht & van Brummelen 1998).

Currently, emphasis is on the study of biomarkers (or bioreporters), which combine high specificity and high sensitivity making use, for example, of the molecular genetic response to specific inorganic and organic contaminants (e.g. Klimowski *et al.* 1996; Murk *et al.* 1996).

Integration of monitoring methods could, however, especially refer to the integration or tiered combination of chemical analysis and biomarkers. In this case the first tier could, again, comprise the use of a low-level biomarker from Table 9.4, but now in combination with specific and sensitive chemical analysis. Other methods for the combination of chemical and biological detection are available. To aid industry and consultants in the identification of toxic compounds present in industrial process and wastewater, a technique called Toxicity Identification Evaluation (TIE) has been developed (see, e.g., Deanovic *et al.* 1999). The method combines physicochemical manipulation of samples with toxicity testing followed by sophisticated chemical analyses. The process consists of three phases: (1) a study of the physicochemical nature of the constituents which are causing the toxicity; (2) isolation of the toxic constituents followed by chemical analyses and identification of the causative agents; and (3) confirmation of the suspected toxicants. Successful application of the TIE process requires an understanding of chemistry and toxicology, as well as a thorough knowledge of mass spectrometry including interpretation of mass spectra from first

Table 9.4 Biomarkers for different classes of pollutants, which are nominated to be included in the safety net (extended from Shugart (1993) and Peakall (1992)).

Environmental pollutant	Biomarker	Temporal occurrence	Sensitivity	Reliability index
Toxic metals	DNA integrity	early/middle	±	s
	Metallothioneins	early	+	s, d
	ALAD inhibition	early	+	s, d, p
	Porphyrin profile	middle	±	s
	Immune response	middle/late	+	s
	Levels of serum enzymes	middle	±	s
	Stress protein	early	±	s
	Neutral red uptake/retention	early	+	s
PAHs	DNA/haemoglobin adducts	early	±	s, d, p
	DNA integrity	early/middle	±	s
	MFO induction	early	++	s, d
	Immune response	middle/late	+	s
	Stress protein	early	±	s
	Neutral red uptake/retention	early	+	s
PHAHs	DNA/haemoglobin adducts	early	±	s, d, p
	DNA integrity	early/middle	±	s
	MFO induction	early	++	s, d
	Porphyrin profile	middle	+	s
	Retinol changes	early	+	s
	Immune response	middle/late	+	s
	Stress protein	early	±	s
	Neutral red uptake/retention	early	+	s
OPs	ACHE inhibition	early	+	s, d, p
	Neuroesterases inhibition	early	+	s, d, p
	DNA integrity	early/middle	±	s
	Enzyme profile	middle	±	s
	Immune response	middle/late	+	s

s = signal of potential problem; d = definitive indicator of type or class of pollutant; p = predictive indicator of long-term adverse effect.

principles. The identification of toxicants through the application of the TIE process has resulted in defining site-specific analytical requirements for industry.

Other types of integration include the combined use of structure–activity relationships (see below) and toxicity-based chemical fractionation techniques in order to allow for compound specific identification (Kosian *et al.* 1998).

The case of MCPA

MCPA, 4-chloro–2-methylphenoxy acetic acid, has been applied as a chlorinated phenoxy carboxylic herbicide for many years in The Netherlands (Teunissen-Ordelman & Schrap 1996) and other countries (Caux *et al.* 1995). MCPA is used world-wide and is, for example, among the top ten herbicides sold in Canada. It has a systemic effect and is used to control a large range of broadleaved weeds in agricultural and non-crop lands.

Similar pesticides, such as 2,4-D and 2,4,5-T, have been shown to contain extremely toxic impurities (chlorinated dioxins), which are formed during their production (Kimbrough & Grandjean 1989; Milnes 1971). Chlorinated dioxins have not been reported to be present in the end product of MCPA. Impurities include isomers of reaction products originating from the series of chemical reactions leading, ultimately, to MCPA (Du Pont 1995). In the traditional way of preparing MCPA, 2-methyl phenol is firstly chlorinated via sulfuryl chloride at 30–40°C. This step results in a mixture of 93% 4-chloro- and 6% 6-chloro–2-methyl phenol isomers. This mixture is subsequently distilled in order to remove the unwanted 6-chloro isomer, leading to a 4-chloro isomer of 97–98% purity. The latter is subjected to a condensation step with chloroacetic acid under alkaline conditions and by refluxing. The resulting product is extracted with organic solvent in order to remove unreacted 4-chloro isomer. After addition of a strong inorganic acid MCPA precipitates.

In a pilot project, prediction of the formation of chlorinated isomers from the chlorination of 2-methyl phenol was studied (Kwast, de Voogt & Govers, in preparation). This type of prediction studies is completely new in environmental chemistry and has not been applied in the design step of agrochemical production processes until now. A first assumption has to be made about the thermodynamic equilibrium or kinetic character of the reaction mechanism, known to be an electrophilic aromatic substitution with a proposed transition state structure (shown in Fig. 9.4). The knowledge of the transition state is essential if a kinetic character for the mechanism is assumed. For a thermodynamic equilibrium approach only an hypothesis on starting reactants and final products is required. Additional assumptions, not specified here, have to be made on the degree of irreversibility of the reaction, and its catalyst and solvent conditions. The reaction (rate or equilibrium) constants, or their ratio in case of the prediction of percentages of isomer yields, could be calculated according to existing models using the available software of THERM and MOPAC.

In addition to the possible formation of the 4-chloro and 6-chloro isomers, the formation of the 3-chloro and 5-chloro isomers, not detected experimentally, was predicted. The results for the kinetic model, which turned out to agree closest to experimental yields, are given in Table 9.5.

Table 9.5 Isomer yield percentages of chlorinated methyl phenols calculated by a kinetic model (Kwast, de Voogt & Govers, in preparation; experimental data from Watson (1976)).

Isomer	% formed (calculated)	% formed (experimental)
3-Chloro-2-methyl phenol	6.0 E-08	Not detected
4-Chloro-2-methyl phenol	99.31	90
5-Chloro-2-methyl phenol	7.0 E-08	Not detected
6-Chloro-2-methyl phenol	0.69	10

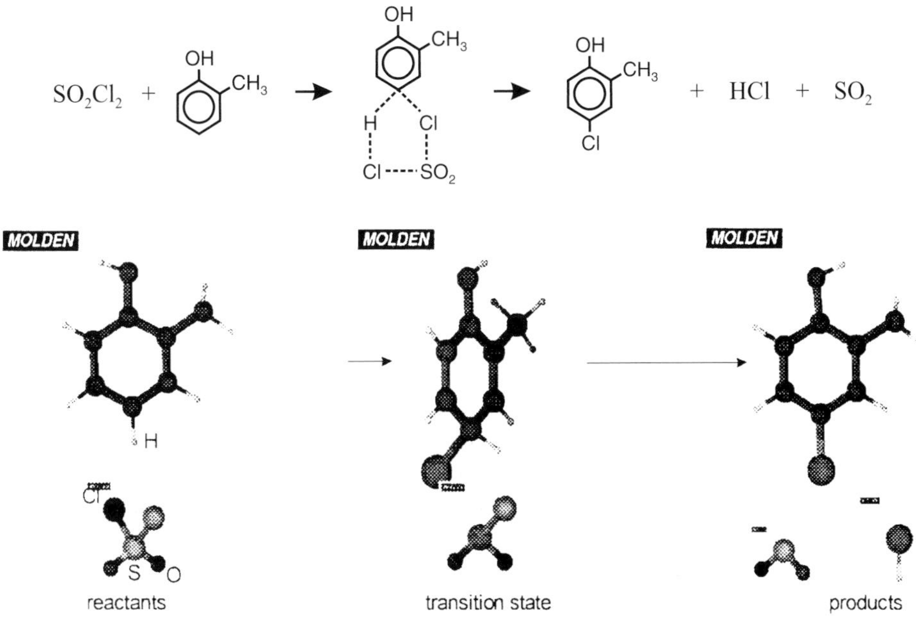

Fig. 9.4 Transition state of the chlorination of 2-methyl phenol as proposed and modelled.

The results are preliminary but promising. Future developments in the design of production processes, contributing in an early stage to an economically and environmentally saving SNOB (RP0, see Fig. 9.2) decision within the company, may be expected. Currently, companies limit their efforts completely to experimental detection of impurities during production, which is also the main feature of RP1 registration with respect to impurities.

Data to be used for fate modelling of MCPA in RP0 and RP1 procedures are only partially available (see Table 9.2; Teunissen-Ordelman & Schrap 1996). In particular, reliable data for (bio-)degradation rate constants are scarcely available. Data refer to degradation in a bubble reactor (Hinteregger & Streichsbier 1999), the complex influence of nutrient conditions on biodegradation (Vink *et al.* 1999) and *in situ* biodegradation in a polluted aquifer (Zipper *et al.* 1998). The compound can undergo biodegradation under aerobic conditions (Caux *et al.* 1995). No fate modelling of impurities or metabolites is available.

With respect to *ex post* (RP2) detection of MCPA and its impurities and metabolites in the environment, the following data were collected. In The Netherlands MCPA is applied on a level of about 200 ton/year and was detected by chemical analysis in surface waters (Teunissen-Ordelman & Schrap 1996). No measures with respect to registration policy are known. In Canada, MCPA has been detected in surface waters at levels varying between 0.00003 and 0.013 mg/L, and at relatively high levels in some groundwater (1.0 mg/L). It has also been reported to affect organisms such as the diatom *Navicula pelliculosa* at levels as low as of 0.026 mg/L, and the beagle dog at concentrations of 0.75 mg/kg/d. The development of the Canadian Water Quality Guidelines for MCPA, which are numerical concentrations of MCPA designed to protect freshwater and marine water life, livestock, and crops, is based on these data (Caux *et al.* 1995).

Currently Mixed Function Oxidase (MFO) and DNA Adduct Formation and other biomarkers can be applied for MCPA (Camatini *et al.* 1998; 1996). In addition bioluminescent whole-cell bioreporters were used recently for the detection of the related compound of 2,4-D (Hay *et al.* 2000). Results demonstrate that cultured cells represent a rapid, controlled and useful method to test pesticides both individually and in combination. In the production of drinking water from river water, the quality of the raw water is extremely important. For this reason the Water Transport Company Rhine-Kennemerland (WRK) in The Netherlands operates, among others, an early warning system. This system was found to be inappropriate for the detection of certain chlorophenoxy herbicides (e.g., 2,4-D and MCPA) that are occasionally found in the river Rhine in increased levels (above the drinking water standard of 0.1 µg/L). Commercially available immunoassay kits were evaluated for use in pre-screening. With some modifications to adapt the working range and to enhance the sensitivity of the kits, they were found to be applicable for early warning. Comparison with a gas chromatography-mass spectrometry (GC-MS) reference chemical analysis revealed no false negative results and a rate of false positive results of about 10%. However, it should be taken into account that matrix effects may affect the results. A daily sampling frequency combined with an analysis frequency based on the actual flow in the river allowed for a periodic analysis. This proved cost effective and also permitted timely availability of the results (Meulenberg & Stoks 1995).

The case of the monoterpenoids

One group of compounds of current interest as an alternative for chlorinated organics is the monoterpenoids, with the known representatives of menthol, citronellal, carvone, thymol and camphor (see Fig. 9.5).

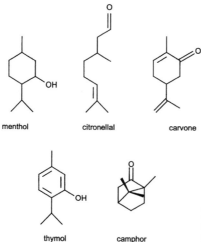

menthol   citronellal   carvone

thymol   camphor

Fig. 9.5   Some monoterpenoid compounds to be used as biomass-based active ingredients.

Several of these non-chlorinated compounds are already used as repellents or pesticides against, for example, the house fly, moths, tobacco cutworm, spruce and larch beetles, and mites (Tambach 2000). These compounds could be alternatives to chlorinated compounds such as chlorpyriphos. Though toxic to non-target organisms such as honey bees and fish, their toxicity is moderate. Moreover they can be produced from plants as one form of biomass-based production. Finally, being natural products, biodegradation could be rapid and complete under environmental conditions. Yet biodegradation should be time-tailored in order to keep the active ingredient's effects. However, biodegradation data of monoterpenoids are largely lacking as are other properties to be known as input data of fate modelling and risk assessment. Prediction methods for biodegradation are under investigation elsewhere (Boethling 1996), although not being applied to monoterpenoids. In addition, no impurity or metabolite data are available, although it is known that plants often produce complex mixtures of structurally related monoterpenoids. Only for camphor and related compounds bacterial hydroxylation and other metabolisation steps are known or can be hypothesised and modelled in a proper way (Paulsen & Ornstein 1992). A key factor in biodegradation is the rate of the determining, (slowest) step of the overall process. Apart from the proper biochemical reactivity, the processes determining the availability for reactions are important such as: desorption from soil and sediment particles to water, which is often the phase to enable uptake by organisms; permeation of organismal membranes, diffusion to degrading enzymes and binding onto the active site of the enzyme prior the reaction (degradation). In order to develop QSAR equations for biodegradation, independent variables (molecular descriptors) are being developed in our department. They are describing, at least, the binding of the monoterpenoid in this active site cavity within the enzyme and the reactivity of enzyme (Iron) and monoterpenoid atoms with respect to each other (see Fig. 9.6). Once these data are available, which holds for the pertinent P450Cam enzyme, reactivity can be predicted, similar to that of MCPA (see above), using available (HYPERCHEM) software.

The development of biodegradation rate constant prediction methods is intended to be included in agrochemical design (see Fig. 9.2). It could be implemented easily in RP0 decision. No data are available on RP1 registration with respect to impurities and biodegradation. Monoterpenoids are not included in current RP2 monitoring in The Netherlands. Maybe this exclusion will be warranted by future results of studies with respect to chemical analysis of natural monoterpenoid mixtures, biodegradation experiments, biodegradation prediction and environmental analysis surveys. In that case this part of the proposed safety net could be reduced substantially for this type of biomass-based pesticides.

Discussion and conclusions

The pesticide safety net proposed has the following potential advantages over existing strategies.

It considers the complete life cycle of a pesticide compound and the societal groups

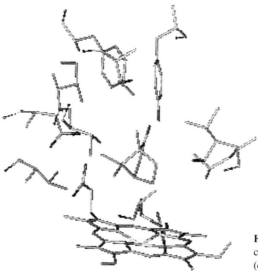

Fig. 9.6 P450Cam enzyme cavity including a monoterpenoid (central molecular part).

connected to stages of this life cycle: a main pre-condition for, e.g., Constructive Technology Assessment (not treated in this paper) and balanced societal policies. Related to this, the accessibility to scientific data for both the general public and research institutions could be improved in this way.

The safety net emphasises and improves three main decision steps ('registration procedures', RP). Firstly, it supports company strategies (RP0) for the economically effective design of chemicals not harmful to the environment. Improvements may especially be found in the prediction of the formation of impurities during production and transformation of parent pesticides in the environment itself by (bio-)degradation and metabolisation.

Secondly, it also improves governmental RP1 registration procedures by including more thoroughly the prediction of impurity formation, environmental degradation and metabolisation of parent compounds.

Thirdly, it strengthens RP2 feedback (*ex post* assessment) by detection of unexpected parent pesticides, their impurities and metabolites, which could lead to more effective RP0 and RP1 *ex ante* decisions. Moreover, the need for RP2 strengthening could be reduced by tailoring the (complete) biodegradability of the compound by the admission and registration of 'biomass-based' chemicals.

In order to give insight in scientific pre-conditions in the area of *ex ante* risk assessment (prediction, RP0, RP1) and *ex post* detection of compounds in the environment (RP2), the availability and urge of methods was reviewed. The following conclusions are drawn:

(1) The development of RP0 and RP1 impurity, metabolisation and biodegradability prediction methods (QSAR) is in a preliminary stage and should be stimulated because they are key parameters in the assessment. The same holds for several (other) input parameters for fate models. The example of the pesticide of MCPA demonstrated the potentialities (no need for experimental

determination) and limitation for the prediction of impurity formation during industrial production. The case of 'biomass-based' monoterpenoids illustrated the difficulties met in the prediction of biodegradation and the ways to solve these.

(2) Available fate models for the RP0 and RP1 prediction of environmental levels should be validated and improved, based on field data and laboratory experiments. No fate modelling was found in the literature in the cases of MCPA and monoterpenoids. This was, at least partially, caused by the lack of experimental input data and reliable prediction methods for these.

(3) Chemical analysis should be integrated with biological detection methods in tiered approaches for RP2 strategies such as Toxicity Identification Evaluation (TIE). Biological methods could be sensitive, rapid and cost effective. In addition, they may give simultaneous information on both presence and effects of compounds. Opportunities are available, but selectivity and knowledge of mechanisms are poor. In particular, biological methods should be improved with respect to criteria relevant to a safety net. Examples are scarce, but available, mainly for aquatic ecosystems. In the case of MCPA chemical detection results turned out to lead political feedback in RP1 in Canada. In The Netherlands they did not. Biomarker detection turned out to be available for MCPA and related compounds. However, in one example they were not appropriate as an early warning system, because of poor detection limits. This is in contrast to commercially available and adapted immunoassay kits. Monoterpenoid compounds, used for example as insect repellent, turned out not to be monitored at all in The Netherlands.

Finally, we would like to emphasise that multidisciplinary (chemical and biological) research could improve both prediction and detection methods and their integrated inclusion in a safety net.

References

Adams, J. (1995) *Risk*. UCL Press, London.

Adriaanse, P.I. (1996) *Fate of pesticides in field ditches: the TOXSWA model*. Report 90 Agricultural Research Department. DLO Winand Staring Centre, Wageningen.

Boesten, J.J.T.I & van der Linden, A.M.A. (1991) Modelling the influence of sorption and transformation on pesticide leaching and persistence. *Journal of Environmental Quality*, 20, 425–35.

Boethling, R.S. (1996) Designing biodegradable chemicals. In: *Designing Safer Chemicals: Green Chemistry for Pollution Prevention* (eds S.C. de Vito & R.L. Garrett), ACS Symposium Series no 640, pp. 156–171. ACS, Washington, DC.

Bolton, R. & de la Mare, P.B.D. (1967) Sulphurylchloride as an electrophile for aromatic substitution. *Journal of the Chemical Society*, B1967, 1044–6.

Camatini, M., Bonfanti, P. & Colombo, A. (1998) Molecular approaches to evaluate pollutants. *Chemosphere*, 37, 2717–38.

Camatini, M., Colombo, A. & Bonfanti, P. (1996) In vitro biological systems as models to evaluate the toxicity of pesticides. *International Journal of Environmental Analytical Chemistry*, 65, 153–67.

Caux, P.Y., Kent, R.A. & Bergeron, V. (1995) Environmental fate and effects of MCPA: A Canadian perspective. *Critical Reviews in Environmental Science and Technology*, 25, 313–76.

CML & TNO (1995) *Een Chloorbalans voor Nederland.* TNO rapporten STB/95/40-I-III, Apeldoorn/Leiden.

Cowan, C.E., Mackay, D., Feytel, T.C.J. *et al.* (1995) *The Multimedia Fate Model: A Vital Tool for Predicting the Fate of Chemicals.* SETAC Press, Pensacola, USA.

Deanovic, L., Connor, V.M., Knight, A.W. & Maier, K.J. (1999) The use of bioassays and toxicity identification evaluation (TIE) procedures to assess recovery and effectiveness of remedial activities in a mine drainage-impacted stream system. *Archives of Environmental Contamination and Toxicology*, 36, 21–27.

DiGiulio, R.T., Habig, C. & Gallaghar, E.P. (1993) Effects of black rock harbor sediments on indexes of biotransformation, oxidative stress, and DNA integrity in Channel Catfish. *Aquatic Toxicology*, 26, 1–22.

van Dijk, H., van Pul, A.W.J. & de Voogt, P. (1999) *Fate of Pesticides in the Atmosphere.* Kluwer Academic Publishers, Dordrecht.

DTO (1997) *DTO Sleutel Chemie; zon en biomassa: bronnen van de toekomst.* Interdepartmental Research Programme, Sustainable Technological Development [DTO]. ten Hagen & Stam, Den Haag [in Dutch].

Du Pont, J. (1995) *Vorming van milieuverontreinigende nevenproducten in de industrieele productie van gechloreerde fenoxycarbonzuren.* Research report, Universiteit van Amsterdam, Amsterdam.

ECB [European Chemicals Bureau] (1997) *EUSES, the European Union System for the Evaluation of Chemicals.* Joint Research Centre European Commission, EUR 17308 EN. ECB, Ispra.

Gelsomino, A., Petrovicova, B., Tiburtini, S., Magnani, E. & Felici, M. (1997) Multi-residue analysis of pesticides in fruits and vegetables by gel permeation chromatography followed by gas chromatography with electron capture and mass spectrometric detection. *Journal of Chromatography*, A782, 105–122.

van Gestel, C.A.M. & van Brummelen, T.C. (1996) Incorporation of the biomarker concept in ecotoxicology calls for a redefinition of terms. *Ecotoxicology*, 5, 217–25.

Govers, H.A.J. (1997) *Prevention of Risks of Organic Micropollutants (PROMPT): A multidisciplinary development of a safety net for organochlorine pesticides*, Intermediate report. University of Amsterdam, Amsterdam.

Harrison, R.M., de Mora, S.J., Rapsomanikis, S., &Johnston, W.R. (1993) *Introductory Chemistry for the Environmental Sciences.* Cambridge University Press, Cambridge.

Hay, R.G., Rice, J.F. & Applegate, B.M. (2000) A bioluminescent whole-cell reporter for detection of 2,4-dichlorophenoxyacetic acid and 2,4-dichlorophenol in soil. *Applied Environmental Microbiology*, 66, 4589–94.

Hinteregger, C. & Streichsbier, F. (1999) Continuous biodegradation of phenoxyalkanoate herbicides by *Sphingomonas herbicidovorans* MH in a PU-supplied bubble reactor. *Acta Biotechnologica*, 19, 279–92.

Karcher, W. & Devillers, J. (1990) *Practical Applications of Quantitative Structure–Activity Relationships (QSAR) in Environmental Chemistry and Toxicology.* Kluwer Academic Publishers, Dordrecht.

Kimbrough, R.D. & Grandjean, P. (1989) Occupational Exposure. In: *Halogenated biphenyls, terphenyls, naphthalenes, dibenzodioxins and related products* (eds R.D. Kimbrough & A.A. Jensen), pp. 485–507. Elsevier, Amsterdam.

Klimowski, L., Rayms-Keller, A., Olson, K.E. *et al.* (1996) Inducibility of a molecular bio-reporter system by heavy metals. *Environmental Toxicology and Chemistry*, 15, 85–91.

de Knecht, J.A. & van Brummelen, T.C. (1998) *Biological assessment of the presence and effects of new and unknown organic contaminants in the environment.* Research report, Vrije Universiteit, Amsterdam.

Kosian, P.A., Makynen, E.A., Monson, P.D. *et al.* (1998) Application of toxicity-based fractionation techniques and structure-activity relationship models for the identification of phototoxic polycyclic aromatic hydrocarbons in sediment pore water. *Environmental Toxicology and Chemistry*, 17, 1021–33.

Kwast, O., de Voogt, P. & Govers, H.A.J. [in preparation] Prediction of isomer ratios in the production of MCPA.

van Leeuwen, C.J. & Hermens, J.L.M. (1995) *Risk Assessment of Chemicals: An Introduction.* Kluwer Academic Publishers, Dordrecht.

Leistra, M., van der Linden, A.M.A., Boesten, J.J.T.I., Tiktak, A. & van den Berg, F. (2000) PEARL model for pesticide behaviour and emissions in soil–plant systems. Description of processes. Report 013. Alterra, Wageningen, The Netherlands.

Linders, J.B.H.J., Luttik, R., Knoop, J.M., van de Meent, D. (1990) *Beoordeling van het gedrag van bestrijdingsmiddelen in oppervlaktewater in relatie tot expositie van waterorganismen.* Report 678611002. RIVM, Bilthoven [in Dutch].

Lynch, M.R. (1995) *Procedures for Assessing the Environmental Fate and Ecotoxicology of Pesticides.* SETAC-Europe, Brussels.

Mackay, D. & Paterson, S. (1990) Fugacity models. In: *Practical Applications of Quantitative Structure–Activity Relationships (QSAR) in Environmental Chemistry and Toxicology* (eds W. Karcher & J. Devillers), pp. 433–460. Kluwer Academic Publishers, Dordrecht.

Mackay, D. & Boethling, R.S. (2000) *Handbook of Property Estimation Methods for Environmental and Health Science.* Lewis Publishers, Boca Raton.

van de Meent, D. (1993) *Simplebox: a generic multimedia fate evaluation model.* Report no. 715501007. RIVM, Bilthoven.

Meulenberg, E.P., & Stoks, P.G. (1995) Water-quality control in the production of drinking-water from river water: The application of immunological techniques for the detection of chlorophenoxy acid herbicides (2,4-D). *Analytica Chimica Acta*, 311, 407–413.

Milnes, M.H. (1971) Formation of 2,3,7,8-Tetrachlorodibenzodioxin by thermal decomposition of sodium 2,4,5-Trichlorophenate. *Nature*, 232, 395–6.

Murk, A.J., Legler, J., Denison, M.S., Giesy, J.P., van de Guchte, C. & Brouwer, A. (1996) Chemical-activated luciferase gene expression (CALUX): a novel *in vitro* bioassay for Ah-receptor active compounds in sediments and pore water. *Fundamental and Applied Toxicology*, 33, 149–60.

Oikari, A., Boon, J.P., Fairbrother, A. *et al.* (1993) Development and validation of biomarkers. In: *Biomarkers* (ed. D.B.P.L.R. Shugart). Springer Verlag, Berlin.

Paulsen, M.D. & Ornstein, R.L. (1992) Predicting product specificity and coupling of cyto-chrome P450Cam. *Journal of Computer-Aided Molecular Design*, 6, 449–60.

Peakall, D.B. (1992) *Animal Biomarkers as Pollution Indicators*. Chapman & Hall, London.

Plummer, E.L. (1990) The application of quantitative design strategies in pesticide discovery. In: *Reviews in Computational Chemistry* (eds K.B. Lipkowitz & D.B. Boyd), pp. 119–168. VCH Publications, New York.

van Roon, A., Govers, H.A.J., Parsons, J.R. & van Weenen, H. (2001) Sustainable chemistry: an analysis of the concept and its integration in education. *International Journal of Sustainability in Higher Education*, 2, 161–79.

Schot, J. (1996) De inzet van constructief technology assessment. *Kennis en Methode*, 3, 265–93.

Shugart, D.B.P.L.R. (1993) *Biomarkers*. Springer Verlag. Berlin.

Tambach, T. (2000) *Evaluation of monoterpene toxicity to target and non-target organisms.* Department of Environmental and Toxicological Chemistry, University of Amsterdam, Amsterdam.

Teunissen-Ordelman, H.G.K. & Schrap, S.M. (1996) *Bestrijdingsmiddelen: Een analyse van de problematiek in het Aquatisch Milieu. Watersysteemverkenningen 1996*. RIZA, Lelystad.

Tiktak, A., van der Linden, F.A. & Swartjes, F.A. (1994) *PESTRAS: A one-dimensional model for assessing, leaching and accumulation of pesticides in soil.* Report no. 715501003. RIVM, Bilthoven, The Netherlands.

Tukker, A. (1998) *Frames in the Toxicity Controversy*. Thesis, Katholieke Universiteit Brabant, Tilburg, The Netherlands.

Vink, J.P.M., Schraa, C. & van der Zee, S.E.A.T. (1999) Nutrient effects on microbial transfor-mation of pesticides in nitrifying surface waters. *Environmental Toxicology*, 14, 329–38.

de Vito, S.C. & Garrett, R.L. (1996) *Designing safer chemicals: green chemistry for pollution prevention.* Symposium Series no 640. ACS, Washington, DC.

Watson, W.D. (1976) The regioselective para chlorination of 2-methyl phenol. *Tetrahedron Letters*, 30, 2591–4.

WCED [World Commission on Environment and Development] (1987) *Our Common Future.* Oxford University Press, Oxford.

Zipper, C., Suter, M.J.F., Haderlein, S.B., Gruhl, M. & Kohler, H.P.E. (1998) Changes in the enantiomeric ratio of R- to S- mecoprop indicate *in situ* biodegradation of this chiral herbicide in a polluted aquifer. *Environmental Science and Technology*, 32, 2020–76.

Chapter 10
Site-Specific Pest Management

Scott M. Swinton

Introduction

Site-specific farming uses information linked to electronic and mechanical technologies in order to enhance spatial input use and the evaluation of crop performance. It offers a spatial complement to biological development-specific technologies, such as threshold-based integrated pest management.

Site-specific pest management (SSPM) today is far from commercial implementation, but a futuristic vision might look like this: imagine a potato farmer who is attempting to manage pests without using genetically-modified organisms and under strict regulations restricting pesticide use. After potato seedlings emerge from soil untreated with pre-emergence pesticides, she traverses the field in a spray rig that optically recognises individual weed species and treats them with the environmentally mildest, most cost-efficient herbicide to which each is susceptible. A few weeks later, based on a freshly made map of Colorado potato beetle infestation, she returns to spot-spray, aware that she has reduced both her spray costs and the rate at which Colorado potato beetle will develop resistance to the insecticide. At season's end, her computer-guided harvester separates out potatoes from unsprayed areas of the field for sale as premium-priced, pesticide-free produce.

Research to enable the elements of this vision is currently under way. But whether SSPM technologies will become feasible and, if so, whether they will become commercially attractive both remain open questions. This chapter reviews the rapid advance of enabling research into site-specific pest management. It begins with an overview of site-specific farming technologies, and moves on to appraise current progress and future directions for a range of crop pests.

Site-specific farming automates much of the spatial 'tailoring' with which farmers of small fields traditionally manage their fields. It is comprised of four major computer-based technologies (Pierce & Sadler 1997; NRC 1997):

(1) Geographic information systems (GIS)
(2) Positioning systems (notably the satellite-based Global Positioning System, GPS)
(3) Variable-rate control applicators
(4) Automated sensing technologies.

GIS are a class of spatially-referenced databases that allow information to be stored by location. GIS software has the ability to store many 'map layers' of information about such diverse spatial attributes as pest populations, soil pH, annual precipitation, and farm-gate prices. Positioning systems, especially the versatile GPS, provide precise location information. Location information can be used to create data layers for a GIS database or to locate a moving piece of equipment that is using a GIS database. A variable-rate input applicator is just such a piece of equipment; it is driven by an electronic controller that can mechanically adjust pesticide rates, as well as leave selected field areas untreated. Jointly, GIS, GPS and variable-rate input application make site-specific farming possible.

The heavy data demands of these three technologies have created demand for a fourth area, the sensing technologies that can gather spatial data at low marginal cost. The most widely used of these are yield monitors, whose data can be stored in a GIS and displayed as a yield map. The yield map, in turn, can be used for spatial performance monitoring. Other sensing technologies can offer useful data for crop management during the season. Remotely sensed data from satellite imagery or aerial photography can be used to make GIS maps for processing into recommendations for subsequent treatment. Proximate sensors in the farm field can enable immediate treatments, for example spot-spraying of weeds whose leaf geometry or light reflectance patterns have been recognised.

Spatial technologies for weed management

The early applications of site-specific farming were for fertiliser application according to maps of soil nutrient deficiency (e.g., Carr *et al.* 1991; Larson & Robert 1991). They focused on phosphorus and potassium management, because those macro-nutrients are quite immobile. The principle of mobility is relevant to SSPM in general. Fig. 10.1 illustrates a stylized continuum of pests from least mobility to greatest. It ranges from weeds to nematodes and from insects to diseases. Because of their immobility, weeds and (to a much lesser extent) nematodes have been the primary targets of SSPM. For each class of pest, innovations in site-specific pest management have focused partly on how to locate pests and partly on how to control pests that have been so located.

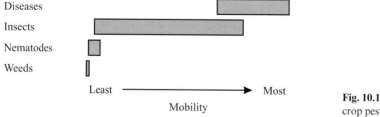

Fig. 10.1   Continuum of crop pest mobility.

Among crop pests, weeds have received the greatest attention from developers of site-specific farming technologies. As was true when Mortensen *et al.* (1998) and Johnson *et al.* (1997) published the first surveys of site-specific weed management, simulation models underline the profitability potential of site-specific weed management (Lindquist *et al.* 1998; Oriade *et al.* 1996). But field research and development of commercial applications have lagged behind. Most efforts in site-specific weed management to date have been directed at characterising weed location, either by mapping or by proximate sensing (Mortensen *et al.* 1998).

Locating weed problems

Mapping weeds for site-specific management poses three important challenges. The first is timeliness. Because weeds in annual crops begin to reduce yield within four to six weeks of weed emergence, required weed control must be prompt (Cousens *et al.* 1987). This implies that weed maps must be made rapidly. The second challenge is to identify weed species, which is important for choosing efficacious herbicides. Unfortunately, the timeliness principle demands that weed maps be made when the weeds are small and hard to recognise. The third challenge is to accomplish the first two at modest cost.

The most reliable (but costly) way to make a multi-species weed map is to rely on intensive sampling by trained weed scouts. A major focus of recent research has been into spatial methods of statistical interpolation that can create a weed map from the set of sample points at which weeds were measured. Remote sensing offers an alternative to in-field scouting, but when weeds are small, remote sensing without ground verification may be unreliable, especially at distinguishing species. Some farmers use binary indicator switches available with most yield monitors digitally to mark field areas where weedy patches exist. Such maps can be quite useful for spot-spraying of perennial weeds prior to the next season's crop. However, their usefulness for control of annual weeds is reduced by uncertainty about whether the same weed species will emerge the next season, given the importance of environmental factors governing the species mix of weed seeds that germinate (Forcella 1992).

Efforts to map weeds using human scouts have chiefly employed grid sampling. Several of these build on previous research demonstrating that the spatial distribution of most weed species is patchy (Mortensen *et al.* 1993, Wiles *et al.* 1992). The most direct approach to grid-based mapping is to attempt to determine the minimum grid density required to get a fair approximation of spatial weed distribution. The geo-statistical method of kriging has been used to interpolate spatial patterns between sampled points. However, two studies found interpolation to be reliable only at sampling densities that were high enough to be prohibitively costly (Clay *et al.* 1999; Dammer *et al.* 1999). At a lower sampling density (roughly 2 ha^{-1}), closer to commercial scale, Wyse-Pester *et al.* (1999) found no spatial correlation for two weed species, although higher densities produced correlations for one of them.

In order to reduce weed sampling costs by substituting soil sample information, some authors have investigated the spatial correlation between soil properties and weed presence (Heisel *et al.* 1999a; Nordmeyer & Dunker 1999). They found significant correlations between the spatial distribution of certain weeds and certain soil characteristics. However, although the prediction variance of weed maps could be improved, all research was conducted with densities of 4–18 plots ha^{-1}, well above the one plot ha^{-1} norm among commercial scouts in the United States. Dieleman *et al.*

(1999) used elevation in addition to soil characteristics to predict the probability of weed presence in a separate field from the one where data were collected. Modest success at predicting the more common weed species suggested that this tool could be used not to replace weed scouting, but rather to assist in directing scouts where weed infestations are most likely.

Another approach to contain sampling costs is to rely on previous data. Hausler & Nordmeyer (1999) found significant inter-annual rank correlation of weed spatial distribution patterns in northern Germany. Although the correlation coefficients were not always positive, evidence of dynamic correlation suggests that weed maps may have some value for control of the subsequent year's weeds, including maps made with crop yield monitor indicator variables.

The use of sensing technologies represents the most active area of research aiming to reduce the cost of data collection on weed location. Among the sensing technologies, the important distinction is between remote sensing (from aircraft or satellites) and proximate (in-field) sensing. Remote sensing is used to make maps from which managers can act, while proximate sensing tends to connect directly to a variable rate herbicide application system that can act on weeds sensed.

Remote sensing offers an automated way to develop weed maps that can be used to guide spot-spraying or to inform field scouts where to seek out economically threatening weed populations. Because the timeliness principle dictates that weeds must be recognised while still small, the first challenge is to distinguish them from the soil in images taken from aircraft (or eventually even satellites). In the face of rapidly improving aerial photography and image processing, research is actively under way on using multispectral imaging and various analysis algorithms, (1) to distinguish weeds from soil, (2) to identify weed types (e.g. grass versus broadleaf) and (3) to identify individual weed species (Lamb *et al.* 1999; Lippert & Wolak 1999; Medlin *et al.* 2000; Varner *et al.* 2000).

Results to date on remote sensing of weed populations suggest that image resolution is key. At resolutions less than 0.5 m, it was difficult to distinguish moderate populations of a grass weed (under 17 wild oat weeds m^{-2}) from soil in an Australian triticale field (Lamb *et al.* 1999). Results are more encouraging for the larger weeds that create economic problems in soybean fields, where recognition rates exceeded 75% of weedy patches exceeding individual species threshold values (Lippert & Wolak 1999; Medlin *et al.* 2000; Varner *et al.* 2000). However, Medlin *et al.* (2000) encountered problems with falsely identifying weed problems in areas that in fact were weed-free. At present, the consensus view appears to be that remote sensing requires ground verification by field scouts in order to be a reliable management tool, implying that remote sensing has not yet achieved its potential to substitute for manual data collection.

Attaining high-image resolution is easier (albeit more costly) by proximate sensing. Proximate image acquisition is the first step toward real-time image processing. Research in this field has focused on either shape recognition or spectral reflectance for distinguishing weeds from background soil or crop and for identifying weed species. Woebbecke *et al.* (1995b) applied image analysis to the geometry of young weed plants, finding that weed shape features evolve rapidly as young plants progress from seedling emergence to development of true leaves. This suggests the need to develop large digital libraries of shape parameters even for a single species. In a related study, the authors evaluated the spectral reflectance characteristics of different weed species (Woebbecke *et al.* 1995a). More recent studies from Europe have combined plant geometry and spectral reflectance to identify different weed species with modest success (Andreasen *et al.* 1997; Martin-Chefson *et al.* 1999). Researchers continue to report high costs and high rates of mistaken identities, suggesting that automated image analysis remains far from commercial readiness.

Treating weed problems

Variable rate herbicide sprayers have been developed that can apply higher rates to field patches where weeds have been mapped. Simpler patch sprayers turn off for weed-free zones. The systems that are currently operational use weed maps or treatment maps based on data collected previous to the field pass at which herbicide is applied (Stafford & Miller 1993; Williams *et al.* 1999). They target post-emergence weed control, either via a tractor-drawn spray rig (Stafford & Miller 1993; Williams *et al.* 1999) or else using irrigation water as the carrier in a centre-pivot or linear-move irrigation system (Eberlein *et al.* 2000).

For weed control prior to crop emergence, recommended herbicide rates are influenced by soil characteristics such as organic matter and soil texture. These attributes can vary within farm fields, opening the door for variable rate herbicide incorporation into the soil. Where herbicide rates have been varied with soil characteristics, as with alachlor in maize, there appears to be potential for reduced chemical use (Khakural *et al.* 1995). However, little research has been conducted on this topic to date (Johnson *et al.* 1997). Research is also under way into variable rate soil-applied herbicide spraying, based on soil organic matter and pH levels.

The logical culmination of timely, accurate, automated weed control is via systems that link machine vision with weed recognition and real-time treatment. Such systems are only in the experimental phase at present. One robotic system for weed control in tomato (Lee *et al.* 1999) takes images, processes them to recognise weeds, and applies herbicide where needed, travelling at 1.20 km/hr. However, the authors concede that the system still leaves much to be desired, with outdoor tests resulting in 24% of tomato plants incorrectly identified and 52% of weeds not sprayed. Key problem areas are weed and crop recognition algorithms and spray accuracy.

Impacts on profitability and environmental impact

Results on the profitability and environmental impact of site-specific weed management remain more heavily based in simulations and hypothetical extrapolations than in experimental data. Several simulated studies of patchy weed distributions have found that spot-spraying of weed patches would lead to significant reductions in herbicide use (Chancellor & Goronea 1994; Johnson *et al.* 1995; Lindquist *et al.* 1998; Medlin & Shaw 2000; Oriade *et al.* 1996). Results on the likely profitability of site-specific weed management are uneven. Certain studies were biased toward finding profitable results by focusing on potential reduced costs from less herbicide spraying while ignoring the increased capital cost of variable-rate application equipment and the increased variable cost of information processing (Lindquist *et al.* 1998; Medlin & Shaw 2000; Oriade *et al.* 1996). With that said, Oriade *et al.* (1996) still found that likely profitability and appeal to producers would depend very much on field characteristics and producer attitude toward the risk of incomplete weed control. Those simulation studies that used fairly complete cost accounting found that profitability was by no means assured, depending on equipment performance (including weed identification and spray targeting) and weed patchiness (Audsley 1993; Bennett & Pannell 1998).

Only a handful of very recent studies have evaluated the effect of site-specific weed management on real fields, all in Europe. In German fields of maize, sugarbeet and wheat, Gerhards *et al.* (1999) were able to reduce herbicide use by half while maintaining effective weed control below thresholds for economic damage. A Danish research team achieved similar herbicide reduction with site-specific spraying using mapped weed data linked to software for determining site-specific weed density thresholds in barley (Heisel *et al.* 1999b). In a maize–sugarbeet rotation, Williams *et al.* (1999) found that uncontrolled weeds were most common not in unsprayed areas, but rather in sprayed areas where more weeds emerged later. They further observed that the number of weed seeds in the soil is high enough that site-specific weed spraying in one year has little effect on weed seedling emergence the following year.

Spatial technologies for management of nematodes, insects and plant diseases

Nematodes

Plant-parasitic nematodes are tiny soil-borne worms that can cause serious crop yield loss by parasitising plant roots and shoots. Due to their limited mobility, parasitic nematodes are also the subject of nascent research into site-specific management. Published research so far has focused on mapping root nematode presence in field crops such as maize (Wyse-Pester *et al.* 1999) and potato (Evans *et al.* 1999; Stafford *et al.* 2000). Both research projects mapped nematode populations by grid soil sampling followed by spatial interpolation between sample points. In maize (Colorado, USA), spatial dependence was not detected with soil samples taken on an 80 m grid; by contrast, in potato (United Kingdom), spatial dependence was present in samples taken on a 20 m grid. In potato, nematode populations after harvest were high enough that site-specific nematicide management seemed less appropriate than rotation out of potato (Evans *et al.* 1999). In one field that had been out of potato and had low levels of potato cyst nematode, only 10% of the field was above the threshold level of 1 nematode egg g^{-1} soil (Stafford *et al.* 2000). However, it was unclear whether the cost of sampling and patch spraying might have overcome the cost of uniform nematicide application even under those conditions.

Insects

Their mobility makes site-specific management of insects considerably more challenging than less mobile species. Whereas weeds and nematodes can be found in a fairly narrow layer just above or just below the soil surface, flying insects occupy a larger area of three-dimensional space. We will focus here on field-level insect management, rather than area-wide insect management, which seeks to eradicate pest insect species within larger geographic regions (see, e.g., Fleisher *et al.* 1997).

Most entomological research at the field level has focused on mapping insect populations and characterising their spatial distributions. As with weeds and nematodes, some efforts employ grid (or block) sampling by insect species. Examples are Colorado potato beetle (*Leptinotarsa decemlineata* (Say)), green peach aphid (*Myzus persicae* (Sulzer)) and potato leafhopper (*Empoasca fabae* (Harris)) in potato (Weisz *et al.* 1995), and western corn rootworm (*Diabrotica vergifera virgifera* (LeConte)), European corn borer (*Ostrinia nubilalis* (Hubner)), western bean cutworm (*Richia albicosta* (Smith)) (Walter & Peairs 1999) and corn rootworm eggs (Wyse-Pester *et al.* 1999), all in the United States. Results for Colorado potato beetle, European corn borer and western bean cutworm adults revealed discernible spatial patterns, with the largest concentrations located close to field borders (Walter & Peairs 1999; Weisz *et al.* 1996).

Due to the high cost of sampling, a separate line of research has explored the use of remotely sensed image analysis for identification of insect problems. Applications to relatively less mobile insect species such as mites and aphids have shown that multispectral imaging can detect mite-induced plant stress in apples and cotton (Fitzgerald *et al.* 2000; Penuelas *et al.* 1995) and aphid damage in wheat (Michels *et al.* 2000; Riedell *et al.* 2000). However, remote sensing cannot always distinguish between plant stress resulting from different sources. One study found it impossible to distinguish between stress due to aphid infestation and stress due to lack of water (Michels *et al.* 2000). Most authors conclude that while remote sensing can be a valuable tool for identifying potential problem areas for scouting, it is not ready to be a substitute for field scouting (e.g., Willers *et al.* 1999).

Due to the limited research into field-level insect mapping, very little site-specific insect treatment has been attempted in annual crops. The leading example to date is the map-based site-specific insecticide application research of Weisz *et al.* (1996) who demonstrated the potential to make reductions of 30–40% in insecticide use for control of Colorado potato beetle in two potato fields in Pennsylvania, USA.

Although proximate sensors have not been applied to insect control in annual crops, commercial sensor applications have developed in fruit tree crops. These so-called 'smart spray' systems use optical vegetation sensing to switch on and off according to the presence of target trees and tree height (Giles *et al.* 1989). Such technologies can reduce spray use in the third dimension in a manner directly analogous to the way that leaving part of an annual crop field saves spray in two dimensions.

Plant diseases

Due to the pervasive presence of plant disease infection agents, research on site-specific disease management has been even more limited. However, some diseases do remain spatially stable once a season is under way, examples being powdery mildew (*Erysiphe graminis*) and *Septoria* spp. These diseases have been the subjects of successful site-specific fungicide management in wheat in Denmark (Bjerre 1999; Secher 1997). The authors found mildew outbreaks most severe along field borders near shelterbelts, while *Septoria* was more closely correlated with crop density. They suggest that future research will need to incorporate climate and temporal crop phenological development: features which have been closer to the forefront of decision support systems for plant pathology during the past twenty years (Travis & Latin 1991).

Environmental assessment of site-specific pest management

Although still at an early stage of development, site-specific pest management holds clear potential to reduce pesticide use. Just as economic thresholds have reduced use *when* pesticides are not needed, so site-specific pest management can reduce use *where* pesticides are not needed. Pesticide use reduction has been documented in numerous simulation studies (Chancellor & Goronea 1994; Johnson *et al.* 1995; Lindquist *et al.* 1998; Medlin & Shaw 2000; Oriade *et al.* 1996), as well as several experiments in farm fields (Gerhards *et al.* 1999; Heisel *et al.* 1999b; Weisz *et al.* 1996). Although reduced pesticide use alone does not ensure reduced environmental impact, it does cut back on the first component of the chain from environmental release to exposure to dose response (Shogren 1990). More focused research on the environmental fate and transport of site-specifically applied pesticides will be required for a definitive analysis of its health and environmental impacts.

For the foreseeable future, the identification and control of weeds are likely to lead the development of technologies for site-specific pest management. In the industrialised countries, herbicide expenditures are the leading agrochemical cost associated with production of most annual crops. Moreover, as plant toxins, herbicides tend to have low human toxicity, and so have been little affected by regulatory restrictions designed to protect human health. Combine these facts with the immobility of weeds on the flat plane of the soil surface, and both the commercial incentive and the technological feasibility of site-specific weed management are present to motivate innovation. The pace of development in the nematode, insect and disease fields will depend upon the mobility (and hence sampling cost) of the respective pest species and upon the continued availability of easily targetable pest control agents.

An important feature of site-specific pest management is that it tends to rely on pest control with chemicals that can be targeted easily. By contrast, most biological pest control agents are difficult to target once released. (Biological agents like *Bacillus thuringiensis* (Bt) are important exceptions.) In a world where banning pesticide active ingredients has become the most common solution to addressing pesticide-related health risks, fewer pesticide formulations are legally available to growers (*cf.* den Hond in this volume). At the same time, experience shows that heavy reliance on a small set of pest controls tends to foster accelerated development of pest resistance to those controls. Colorado potato beetle, for example, has successfully developed resistance to at least three different insecticides used in the United States over the past twenty years. Hence, at the field level, site-specific pest management poses a conundrum: it can reduce pesticide use, yet it relies on the availability of pesticides to be effective. Given the risks of pests developing resistance to heavily used pesticides, SSPM is likely to work best if a variety of pesticides are available to farmers.

Ironically, one motivator of regulatory bans on selected pesticides is the calculated risk of human and environmental exposure to pesticidal toxins. This calculated risk is based on manufacturers' recommended application rates. Yet SSPM inherently reduces field-scale rates by reducing the area over which pesticides are applied.

SSPM's potential to protect consumers from pesticide-related health risks is greatest when government rules focus on pesticide use outcomes, such as residues on food products. There are several ways to accomplish this, including residue taxes and standards (Swinton & Batie 2001). All such 'flexible incentives' would raise the cost of pesticide use without eliminating any but the most dangerous options. In the long run, such output-focused incentives would encourage pesticide product innovation to focus on low-risk, highly targeted, low-rate pesticides (Swinton & Casey 1999). In the best of cases, the new technologies so induced might produce innovation offsets that provide productivity gains to compensate the costs of regulatory compliance (Wossink & de Koeijer in this volume). But past experience shows such beneficial innovation to result only from flexible, outcome-oriented environmental regulation.

Prognosis for adoption of SSPM and pesticide reduction

The futuristic vision of potato farming that opened this chapter is far from reality. Most of the technologies envisioned are currently under research and not commercially available. As some do become available, their adoption will hinge on expected profitability. This, in turn, depends upon (a) efficacy at reducing yield damage (both quantity and quality), and (b) reducing pesticide costs enough to compensate for the added information management costs of SSPM. Little adoption of SSPM practices is likely – in the United States and Europe – before 2005. Even as adoption begins, it is likely to be very partial (only selected practices) and geographically patchy, rather than comprehensive like the picture in the vision.

Adoption of SSPM methods is most likely where pests are relatively immobile – hence with weeds and nematodes. Much will depend on the technical feasibility of accurately predicting or sensing pest location and accurately treating pests found. Timeliness is crucial. Costs and product prices drive profitability. So if sensing technologies became accurate and reliable, they would cut information collection costs dramatically. Where pest controls are expensive or crop products are valuable, or both, adoption is more likely. Finally, government policies can affect the expected profitability of adopting SSPM, for example by cost-sharing the adoption of SSPM practices or increasing penalties for excessive agrochemical use.

The prognosis for adoption of SSPM (and consequent reduction in potential pesticide use) depends on both necessary and sufficient conditions. The key necessary condition is that SSPM technologies must become available commercially. Even in the United States, where site-specific farming is most advanced, very little variable rate spray applicator technology was available by mid-2001. No weed recognition technology was available at that time (with the exception of combine harvester operators mapping perennial weeds with the help of yield monitors).

The key sufficient condition for adoption of SSPM methods is that the added value from increased yield quantity and/or quality must exceed the added net cost of pest control. For nutrient management, this positive benefit:cost ratio has occurred so far only in medium- to high-value crops. It is becoming evident for fertilisers that renewed agronomic research will be needed to redefine site-specific rates.

The same principles apply to pest management: limited, early adoption in high-value crops accompanied by a need for renewed research on spatial pest management. The high cost of monitoring pest populations as they evolve means that SSPM is likely to be adopted only for those pests with simple, easily predicted population dynamics. Much additional research will be required before biologists can predict how selected pest populations will evolve when some areas are left uncontrolled. Until such information becomes available, the adoption of SSPM technologies will remained limited and they will be unable to realise fully their potential to reduce unneeded pesticide use.

Acknowledgement

For helpful comments, the author thanks Lori Wiles, Haddish Melakeberhan and Gary R. Van Ee.

References

Andreasen, C., Rudemo, M., Sevestre, S. (1997) Assessment of weed density at an early stage by use of image processing. *Weed Research*, 37, 5–18.

Audsley, E. (1993) Operational research analysis of patch spraying. *Crop Protection*, 12 (March), 111–19.

Bennett, A. & Pannell, D.J. (1998) Economic evaluation of a weed-activated sprayer for herbicide application to patchy weed populations. *Australian Journal of Agricultural Economics*, 42, 389–408.

Bjerre, K.D. (1999) Disease maps and site-specific fungicide application in winter wheat. In: *Precision Agriculture '99*. (ed. J.V. Stafford), pp. 495–504. Sheffield Academic Press, Sheffield.

Carr, P.M., Carlson, G.R., Jacobsen, J.S., Nielsen, G.A., Skogley, E.O. (1991) Farming by soils, not fields: A strategy for increasing fertilizer profitability. *Journal of Production Agriculture*, 4, 57–61.

Chancellor, W.J. & Goronea, M.A. (1994) Effects of spatial variability of nitrogen, moisture, and weeds on the advantages of site-specific applications for wheat. *Transactions of the ASAE*, 37, 717–24.

Clay, S.A., Lems, G.J., Clay, D.E., Forcella, F., Ellsbury, M.M. & Carlson, C.G. (1999) Sampling weed spatial variability on a fieldwide scale. *Weed Science*, 47, 674–81.

Cousens, R., Brain, P., O'Donovan, J.T. & O'Sullivan, P.A. (1987) The use of biologically realistic equations to describe the effects of weed density and relative time of emergence on crop yield. *Weed Science*, 35, 720–25.

Dammer, K.H., Schweigert T. & Wittmann, C. (1999) Probability maps for risk assessment in a patchy weed control. *Precision Agriculture*, 1 (September), 185–98.

Dieleman, J.A., Mortensen, D.A. & Young, L.J. (1999) Predicting within-field weed species occurrence based on field-site attributes. In: *Precision Agriculture '99* (ed. J.V. Stafford), pp. 517–28. Sheffield Academic Press, Sheffield.

Eberlein, C.V., King, B.A. & Guttieri, M.J. (2000) Evaluating an automated irrigation control systems for site-specific herbigation. *Weed Technology*, 14, 182–7.

Evans, K., Webster, R. & Barker, A. *et al.* (1999) Mapping infestations of potato cyst nematodes and the potential for patch treatment with nematicides. In: *Precision Agriculture '99*, (ed. J.V. Stafford), pp. 505–515. Sheffield Academic Press, Sheffield.

Fitzgerald, G.J., Maas, S.J. & deTar, W.R. (2000) Multispectral multitemporal remote sensing for spider mite detection in cotton. In: *5th International Conference on Precision Agriculture*, (ed. P.C. Robert). CD-ROM, American Society of Agronomy, Precision Agriculture Center, University of Minnesota, St Paul, MN.

Fleisher, S.J., Weisz, R., Smilowitz, Z. & Midgarden, D. (1997) Spatial variation in insect populations and site-specific integrated pest management. In: *The state of site-specific management for agriculture* (eds F.J. Pierce & E.J. Sadler), pp. 101–130. American Society of Agronomy, Madison, WI.

Forcella, F. (1992) Prediction of weed seedling densities from buried seed reserves. *Weed Research* 32, 29–38.

Gerhards, R., Sokefeld, M., Timmermann, C., Reichart, S., Kuhbauch, W. & Williams, M.M. II (1999) Results of a four-year study on site-specific herbicide application. In: *Precision Agriculture '99* (ed. J.V. Stafford), pp. 689–97. Sheffield Academic Publishers, Sheffield.

Giles, D.K., Delwiche, M.J. & Dodd, R.B. (1989) Sprayer control by sensing orchard crop characteristics: Orchard architecture and spray liquid savings. *Journal of Agricultural Engineering Research*, 43, 271–89.

Hausler, A. & Nordmeyer, H. (1999) Characterizing spatial and temporal dynamics of weed seedling populations. In: *Precision Agriculture '99*, (ed. J.V. Stafford), pp. 463–472. Sheffield Academic Press, Sheffield.

Heisel, T., Ersboll, A.K. & Andreasen, C. (1999a) Weed mapping with co-kriging using soil properties. *Precision Agriculture*, 1 (January), 39–52.

Heisel, T., Christensen, S. & Walter, A. M. (1999b) Whole-field experiments with site-specific weed management. In: *Precision Agriculture '99* (ed. J.V. Stafford), pp. 759–68. Sheffield Academic Press, Sheffield.

Johnson, G.A., Cardina, J. & Mortensen, D.A. (1997) Site-specific weed management: Current and future directions. In: *The state of site-specific management for agriculture* (eds F.J. Pierce & E.J. Sadler), pp. 131–147. American Society of Agronomy, Madison, WI.

Johnson, G.A., Mortensen, D.A. & Martin, A.R. (1995) A simulation of herbicide use based on weed spatial distribution. *Weed Research*, 35, 197–205.

Khakural, B.R., Robert, P.C. & Koskinen, W.C. (1995) Runoff and leaching of alachlor under conventional and soil-specific management. *Soil Use and Management*, 10, 158–64.

Lamb, D.W., Weedon, M.M. & Rew, L.J. (1999) Evaluating the accuracy of mapping weeds in seedling crops using airborne digital imaging: *Avena* spp. in seedling triticale. *Weed Research*, 39, 481–92.

Larson, W.E. & Robert, P.C. (1991) Farming by soil. In: *Soil management for sustainability*, (eds R. Lal & F.J. Pierce) pp. 103–112. Soil and Water Conservation Society, Ankeny, IA.

Lee, W.S., Slaughter, D.C. & Giles, D.K. (1999) Robotic weed control system for tomatoes. *Precision Agriculture*, 1 (January), 95–113.

Lindquist, J.L., Dieleman, A., Mortensen, D.A., Johnson, G.A. & Wyse-Pester, D.Y. (1998) Economic importance of managing spatially heterogeneous weed populations. *Weed Technology*, 12, 7–13.

Lippert, R.M. & Wolak, F.J. (1999) Weed mapping and assessment of broadcast vs. spot treatment of sicklepod weeds in soybeans. In: *Precision Agriculture '99*, (ed. J.V. Stafford), pp. 223–227. Sheffield Academic Publishers, Sheffield.

Martin-Chefson, L., Chapron, M., Philipp, S., Assemat, L. & Boissard, P. (1999) A two-dimensional method for recognising weeds from multiband image processing. In: *Precision Agriculture '99* (ed. J.V. Stafford), pp. 473–83. Sheffield Academic Publishers, Sheffield.

Medlin, C.R. & Shaw, D.R. (2000) Economic comparison of broadcast and site-specific herbicide applications in non-transgenic and glyphosate-tolerant *Glycine max*. *Weed Science*, 48, 653–61.

Medlin, C.R., Shaw, D.R., Gerard, P.R. & LaMastus, F.E. (2000) Using remote sensing to detect weed infestations in *Glycine max*. *Weed Science*, 48, 393–8.

Michels, G.J., Piccinni, G., Rush, C.M. & Fritts, D.A. (2000) Using infrared transducers to sense greenbug infestation in winter wheat. In: *5th International Conference on Precision Agriculture*, (ed. P.C. Robert). CD-ROM, American Society of Agronomy, Precision Agriculture Center, University of Minnesota, St Paul, MN.

Mortensen, D.A., Dieleman, J.A. & Johnson, G.A. (1998) Weed spatial variation and weed management. In: *Integrated weed and soil management* (eds J.L. Hatfield, D.D. Buhler & B.A. Stewart), pp. 293–309. Ann Arbor Press, Chelsea, MA.

Mortensen, D.A., Johnson, G.A. & Young, L.J. (1993) Weed distribution in agricultural fields. In: *Proceedings of soil specific crop management: A workshop on research and development issues* (eds P.C. Robert, R.H. Rust & W.E. Larson), pp. 113–124. American Society of Agronomy, Crop Science Society of America, and Soil Science Society of America, Madison, WI.

NRC [National Research Council] (1997) *Precision Agriculture in the 21st Century: Geospatial and Information Technologies in Crop Management*. National Academy Press, Washington, DC.

Nordmeyer, H. & Dunker, M. (1999) Variable weed densities and soil properties in a weed mapping concept for patchy weed control. In: *Precision Agriculture '99* (ed. J.V. Stafford), pp. 453–62. Sheffield Academic Press, Sheffield.

Oriade, C.A., King, R.P., Forcella, F. & Gunsolus, J.L. (1996) A bioeconomic analysis of site-specific management for weed control. *Review of Agricultural Economics*, 18, 523–35.

Penuelas, J., Filella, I., Lloret, P., Muñoz, F. & Vilajeliu, M. (1995) Reflectance assessment of mite effects on apple trees. *International Journal of Remote Sensing*, 16, 2727–33.

Pierce, F.J. & Sadler, J.D. (1997) *The State of Site-Specific Management for Agriculture*. Agronomy Society of America, Crop Science Society of America, and Soil Science Society of America, Madison, WI.

Riedell, W.E., Hesler, L.S., Osborne, S.T. & Blackmer, T.M. (2000) Remote sensing of insect damage in wheat. In: *5th International Conference on Precision Agriculture*, (ed. P.C. Robert). CD-ROM, American Society of Agronomy, Precision Agriculture Center, University of Minnesota, St Paul, MN.

Secher, B.J.M. (1997) Site-specific control of diseases in winter wheat. *Aspects of Applied Biology*, 48, 57–64.

Shogren, J.F. (1990) *A Primer on Environmental Risk Analysis*. Center for Agricultural and Rural Development, Iowa State University, Ames, IA.

Stafford, J.V., Evans, K., Barker, A., Halford, P.D. & Russell, M.D (2000) Changes in the within-field spatial distribution of potato cyst nematodes before and after cropping with potatoes and the consequences for modulating nematicide application. *Aspects of Applied Biology*, 59, 1–7.

Stafford, J.V. & Miller, P.C.H. (1993) Spatially selective application of herbicide to cereal crops. *Computers and Electronics in Agriculture*, 9, 217–29.

Swinton, S.M. & Batie, S.S. (2001) FQPA: Pouring out (in?) the risk cup. *Choices*, 16 (1), 14–17.

Swinton, S.M. & Casey, F. (1999) From adoption to innovation of environmental technologies. In: *Flexible Incentives for the Adoption of Environmental Technologies in Agriculture* (eds F. Casey, A. Schmitz, S.M. Swinton & D. Zilberman), pp. 351–60. Kluwer Academic Publishers, Boston.

Travis, J.W. & Latin, R.X. (1991) Development, implementation, and adoption of expert systems in plant pathology. *Annual Review of Phytopathology*, 29, 343–60.

Varner, B.L., Gress, T.A. & Copenhaver, K. (2000) Weed detection in soybeans using hyperspectral and multispectral imagery. In: *5th International Conference on Precision Agriculture* (ed. P.C. Robert). CD-ROM, American Society of Agronomy, Precision Agriculture Center, University of Minnesota, St Paul, MN.

Walter, S.M. & Peairs, F.B. (1999) Spatial variability of trap catches of three insect pests in sprinkler irrigated maize in eastern Colorado. In: *Precision Agriculture '99*, (ed. J.V. Stafford), pp. 229–37. Sheffield Academic Press, Sheffield.

Weisz, R., Fleischer, S. & Smilowitz, Z. (1995) Site-specific integrated pest management for high-value crops: Sample units for map generation using the Colorado potato beetle (Coleoptera: Chrysomelidae) as a model system. *Journal of Economic Entomology*, 88, 1069–80.

Weisz, R., Fleischer, S. & Smilowitz, Z. (1996) Site-specific integrated pest management for high-value crops: Impact on potato management. *Journal of Economic Entomology*, 89, 501–509.

Wiles, L.J., Oliver, G.W., York, A.C., Gold, H.J. & Wilkerson, G.G. (1992) The spatial distribution of broadleaf weeds in North Carolina soybean (*Glycine max*) fields. *Weed Science*, 40, 554–7.

Williams, M.M. II, Gerhards, R. & Mortensen, D.A. (1999) Spatiotemporal outcomes of site-specific weed management in maize. In: *Precision Agriculture '99* (ed. J.V. Stafford), pp. 897–906. Sheffield Academic Press, Sheffield.

Willers, J.L., Seal, M.R. & Luttrell, R.G. (1999) Remote sensing, line-intercept sampling for tarnished plant bugs (Helioptera: Miridae) in mid-south cotton. *Journal of Cotton Science*, 3, 160–70.

Woebbecke, D.M., Meyer, G.E., von Bargen, K. & Mortensen, D.A. (1995a) Color indices for weed identification under various soil, residue, and lighting conditions. *Transactions of the ASAE*, 38, 259–69.

Woebbecke, D.M., Meyer, G.E., von Bargen, K. & Mortensen, D.A. (1995b) Shape features for identifying young weeds using image analysis. *Transactions of the ASAE*, 38, 271–81.

Wyse-Pester, D., Westra, P. & Wiles, L.J. (1999) Spatial sampling and analysis of crop pests in a center pivot corn field. In: *Precision Agriculture '99* (ed. J.V. Stafford), pp. 485–94. Sheffield Academic Press, Sheffield.

Chapter 11
New Biotechnology, Crop Protection and Sustainable Development

Susan Carr

Introduction

The advent of 'new biotechnology' represents a step change in agricultural production as radical as the 'green revolution' of the 1960s. Then, large yield increases were achieved by the breeding of rice and other cereal varieties that produced much higher yields than conventional varieties in response to high inputs of fertiliser, chemical pesticides and irrigation. While hugely successful in terms of agricultural production, at least in the short to medium term, the 'green revolution' had many unanticipated adverse impacts when judged in relation to its social and environmental context. Biotechnology presents a new opportunity to learn from those past lessons and adopt strategies that bring benefits to agriculture, society and the environment: in effect, to contribute to more sustainable development. Whether or not this opportunity will be realised is a fiercely contested question.

Because the topics of biotechnology and its potential to contribute to sustainable development are so fiercely debated, it is as well to start with some definitions. Biotechnology has been defined by Ervin *et al.* (2000) as:

> the scientific or industrial manipulation of life forms (organisms) to produce new products or improve upon existing organisms (plants, animals or microbes), first coined to apply to the interaction of biology and human technology. In recent usage, refers to all parts of the industry that creates, develops and markets a variety of products wilfully manipulated on a molecular and/or cellular level.

In this chapter, the term biotechnology is used mainly in relation to transgenic crop plants: plants that have been manipulated to contain DNA from at least one unrelated organism, including a virus, bacterium, animal or other plant species. However, the term biotechnology also encompasses techniques such as plant tissue culture, embryo transfer and cell fusion.

There are many definitions of the phrase 'sustainable development'. For example, a strategy document issued by the UK government's environment ministry identified the following four components (DETR 1998):

(1) social progress that meets the needs of everyone;

(2)   effective protection of the environment;
(3)   prudent use of natural resources;
(4)   maintenance of high and stable levels of economic growth and employment.

Critics of that document offered an alternative definition that emphasised its dynamic nature, the importance of social justice and equity, and the idea that there are biophysical limits to human activity: 'Sustainable development is a dynamic process which enables all people to realise their potential and to improve their quality of life in ways which simultaneously protect and enhance the Earth's life support systems' (Forum for the Future 1998).

Sustainable development in relation to agriculture has usually been interpreted more narrowly, to mean optimising (or reducing) the use of synthetic pesticides and minimising environmental impact. In the debate about the potential contribution of biotechnology to sustainable development, three distinct perspectives on sustainable agriculture can be detected: a market-based one, a community-based one, and an environmental management-based one (Levidov 2000). People who argue from a market-based perspective hold that increasing the intensity of agricultural production in some areas will release land that can be managed less intensively in other areas (i.e. a dual system, with separate areas devoted to intensive agriculture and to extensive agriculture or habitat conservation). Those who argue from a community-based perspective see sustainable agriculture in terms of local production without chemical inputs (e.g. organic agriculture). Those who argue from an environmental management perspective envisage sustainable agriculture as an integrated system, with the use of management skill to minimise the impact of agriculture on the environment (e.g. integrated crop management systems) (*cf.* Struik & Kropff in this volume). Some analysts suggest that, to achieve agricultural sustainability, the policy goal should be to encourage all three approaches as part of a complex mix of farming systems, to allow maximum flexibility of response to changing circumstances (Tait & Morris 2000).

This chapter examines the debate about biotechnology's potential contribution to sustainable development, variously defined, by seeking answers to three main questions:

(1)   How might biotechnology contribute to, or impede, the adoption of 'sustainable agriculture' strategies?
(2)   What does the evidence available so far suggest about biotechnology's potential contribution to sustainable agriculture?
(3)   How might biotechnology's contribution to sustainable agriculture be promoted?

The next section sets out some of the background to the commercial uptake of crop biotechnology. Then, the third and fourth sections look at the claims being made about the benefits and risks of crop biotechnology. The next two sections examine the evidence so far available about the benefits and risks. The last section discusses the potential of biotechnology to contribute to sustainable agriculture in the light of those claims and that evidence. It concludes with recommendations about the policies that need to be in place if biotechnology's potential to contribute to sustainable development is to be realised. Inevitably, in a broad-ranging review of this nature, not all aspects and examples receive equally critical treatment, so as many references as possible are included to allow readers to examine the sources and form their own independent conclusions.

Background

Research into genetically modified (GM) crops began in the 1980s, and seeds first became available commercially in significant quantities in 1996 when 1.7 million hectares were grown. Five years later, in 2000, the area planted to GM crops had increased 25-fold to 44 million hectares. Four countries accounted for 99% of that area: USA (68%), Argentina (23%), Canada (7%) and China (1%) (James 2000). The change in the area planted to GM crops between 1997 and 2000 by country is shown in Table 11.1.

Table 11.1 Area planted to GM crops by country 1997–2000 (million ha) (James 2000; GeneWatch UK 1999; 2000a; 2001).

	1997	1998	1999	2000
USA	8.1	20.5	28.7	30.3
Argentina	1.4	4.3	6.7	10.0
Canada	1.3	2.8	4.0	3.0
China		<0.1	0.3	0.5
South Africa		<0.1	0.1	0.2
Australia	0.1	0.1	0.1	0.15
Mexico	<0.1	<0.1	<0.1	<0.1
Spain		<0.1	<0.1	<0.1
France		<0.1	<0.1	<0.1
Germany			<0.1	<0.1
Portugal			<0.1	0
Romania			<0.1	<0.1
Ukraine			<0.1	<0.1
Bulgaria				<0.1
Uruguay				<0.1
Total	12	27.8	39.9	44.2

Soybean, maize, oilseed rape and cotton were the first GM crops to be grown commercially on a significant scale. They incorporated genes for herbicide tolerance (to the broad-spectrum herbicides glyphosate or glufosinate) and insect resistance (to certain species of Lepidoptera and Coleoptera, conferred by a gene coding for the toxin produced by the bacterium *Bacillus thuringiensis*, or Bt). The proportion of the global area of GM crops devoted to each of these crop–trait combinations in 2000 is shown in Table 11.2. That year, herbicide-tolerant soybean accounted for more than half (59%) of the total GM-cropped area.

Table 11.2   Proportion of the global area of GM crops devoted to each crop–trait combination in 2000 (James 2000).

	Percentage of global GM-cropped area in 2000
Herbicide-tolerant soybean	59
Bt maize	15
Herbicide-tolerant oilseed rape	6
Herbicide-tolerant maize	5
Herbicide-tolerant cotton	5
Bt and herbicide-tolerant cotton	4
Bt cotton	3
Bt and herbicide-tolerant maize	3

In northern America, the initial rapid rate of increase showed signs of levelling off in 2000. In the US, although the area planted to GM soybean, cotton and oilseed rape continued to increase, the area of GM maize decreased. This decrease may have been due partly to a reduced incidence of the target pest the previous year, leading farmers to think that the extra cost of Bt seed might not be worthwhile, and partly due to uncertainty about the market for GM maize. In Canada, there was a net decrease in the GM crop area in 2000, mainly because of a decrease in the area of GM oilseed rape. This was ascribed partly to poor market prices and partly to the availability of non-GM herbicide-tolerant oilseed rape varieties (James 2000). Of the 11% increase between 1999 and 2000 in the area planted to GM crops world-wide, 84% was in developing countries and only 16% in industrial countries (James 2000).

In the European Union (EU), GM crops began to be approved for commercial uses in 1996 (Table 11.3). GM crops were first grown commercially in the EU in 1998, when Bt maize was planted on 20 000 ha in Spain and 2000 ha in France. These plantings coincided with a period of intense controversy in many EU member states about GM crops and food, triggered by the arrival of imports of GM soybean and maize from the US from late 1996 onwards (for details, see Carr 2000; Levidov *et al.* 2000).

Previously, non-governmental organisations (NGOs) in Europe opposed to GM technology had found it difficult to find a clear focus for a public campaign. Once

Table 11.3   GM commodity crops approved for commercial use in the European Union (GeneWatch UK 1999; 2000b).

Crop	GM trait	Company	Uses	Approval date
Oilseed rape	HT	Plant Genetic Systems	Seed production only	1996
Soybean	HT	Monsanto	Import for food and feed	1996
Maize	HT, IR	Ciba Geigy (Novartis)	Crop, food and feed	1997
Oilseed rape	HT	AgrEvo	Import, seed production	1998
Maize	HT	AgrEvo	Crop	1998
Maize	HT	Monsanto	Import, food and feed	1998
Maize	HT, IR	Northrup King (Novartis)	Import, food and feed	1998

(HT = herbicide-tolerant; IR = insect-resistant)

products arrived on the market, it became easier to attract media and public attention, especially since the first products arrived from the US ('US neo-imperialism') included soybean (which is an ingredient of very many supermarket products), and were not segregated or labelled as 'GM' (so preventing consumers' choice to avoid GM products). The consequent public outcry, even in countries such as Spain and Italy where previously there had been almost no public debate on the subject of GM products, led major food retailers and processors to provide guaranteed GM-free lines to reassure their customers.

Those EU member states (such as Denmark and Austria) that had always expressed reservations about GM crop approvals gained strength for their position from the public resistance. Their concerns about such issues as gene flow from GM crops to wild relatives, the possible spread of antibiotic-resistant genes to gut pathogens, and the wider and longer-term environmental implications of changing farm practices with the adoption of GM crops, became more widely shared among member states to the point where the regulatory approval system ground to a halt in 1998. Some member states banned GM crops that had already received EU-wide market approval. For example, Austria and Luxembourg, and initially Italy, banned the use of the Novartis insect-resistant and herbicide-tolerant maize. France introduced a two-year moratorium on the commercial cultivation of oilseed rape and sugarbeet. In addition, France refused to sign the final consent for two herbicide-tolerant oilseed rape varieties, which it had previously recommended for EU-wide approval and which had otherwise completed their passage through the regulatory procedure. The UK and Denmark negotiated with industry a delay in the introduction of commercial GM crops, until farm-scale trials on some of their potential environmental impacts were completed.

As the debate in Europe intensified, proposals for revising the EU directive governing GM crop approvals (Directive 90/220) became increasingly precautionary. The text finally agreed in 2001 (as Directive 2001/18) included (FoEE 2001a):

- specific references to the precautionary principle;
- clarification that risk assessment must cover indirect, delayed and cumulative adverse effects, including those resulting from changes in use or management;
- a requirement that risk assessment must 'give particular attention to' antibiotic-resistance marker genes, with a view to phasing out those with an adverse impact;
- a requirement for applicants to submit a monitoring plan, recording who will be responsible, to confirm assumptions made in the risk assessment and to identify adverse effects that were not anticipated;
- locations of all GMOs to be recorded on public registers;
- consents to be conditional on labelling of all GM products;
- first-time marketing consents to be limited to ten years; applications for renewal to include the results of monitoring;
- duty on member states to ensure that GMOs are 'traceable' at all stages.

Even so, six member states (Denmark, France, Greece, Italy, Luxembourg and Austria) said they would not consider any further applications to market GM products until questions concerning traceability and labelling had been resolved by new legislation (ENDs Report 2001; FoEE 2001b).

The situation in Europe created considerable uncertainty for growers in North America, especially when Archer Daniels Midland, the largest grain handler in America, announced in 1999 that they would only accept GM crops already approved in the European Union (GeneWatch UK 2000a). Archer Daniels Midland and another large grain handler, AE Stanley, advised US farmers not to plant GM maize in 2001, after a GM maize variety approved only for use as animal feed was found in food products (ENDs Report 2000). Other important US export markets (for example, Japan and South Korea) imposed restrictions, such as labelling requirements, on the import of GM crops. The markets for GM-free and organic products received a boost as a result of consumer resistance to GM products.

The resistance to GM crops among consumers in Europe, and the knock-on effects along the food retailing, processing and production chain, mean that company and national government policy decisions concerning the future of agricultural biotechnology are more open to scrutiny and debate than before. While the controversy about GM in the EU causes uncertainty for biotechnology companies and US exporters, the resulting pause in regulatory approvals also provides opportunities for a fuller exploration of the potential benefits and risks and a chance to influence biotechnology's research and development trajectory.

Claims about potential benefits

Multinational agrochemical companies, in their publicity material, annual reports and in interview comments made by senior managers, stress a number of benefits that biotechnology potentially offers. Commonly expressed views about the ways in which company strategy on biotechnology can contribute to sustainable agriculture and sustainable development are shown in Table 11.4. (Some of the companies mentioned in Table 11.4 have merged since this research was done: e.g. Novartis and Zeneca merged in 2000 to form Syngenta; Rhône-Poulenc and AgrEvo merged in 1999 to form Aventis; *cf.* den Hond in this volume.)

In particular, companies stress the potential for GM crops to reduce the use of insecticides and herbicides, and to allow farmers to switch to herbicides with less environmental impact. They say that in some cases GM crops will result in improved yields, for example by reducing losses due to pest damage. This will allow more food to be grown on the same area, and reduce the need to extend the area under cultivation, so enabling less intensive production or conserving habitat on the remaining land (a view corresponding with the 'market-based' concept of sustainable agriculture). Many companies point to the predicted increase in population in developing countries, arguing that the use of biotechnology will be essential if the resulting increased demand for food is to be met.

In individual interviews, company managers express more nuanced views (Tait *et al.* 2001). For example, some acknowledge that the benefits of the first generation of GM crops may not be widely shared. Some expect that GM crops, especially those

Table 11.4 Commonly expressed views in multinational agrochemical companies about biotechnology's contribution to sustainable agriculture and sustainable development (Tait *et al.* 2001).

Commonly expressed views	Company
Reducing use and impacts of chemical inputs	
• Reducing use and dependence on chemical plant protection products/ substituting chemicals with GM crop technology	Advanta, Novartis
• Making crops tolerant to environmentally sound and easily degradable herbicides/decreasing impacts of spraying toxic chemicals/ avoiding unwanted effects of non-selective pesticide treatments/ helping decrease pesticide load/reducing pesticide residues	AgrEvo, Danisco, KWS, Novartis, Pioneer, Rhône-Poulenc
• Developing plants that need less nitrogen and that absorb nutrients more effectively	Advanta
Reconciling high yields and reduced environmental impact	
• Supporting high-yield agriculture which takes less space and leaves more land for nature/using arable land more efficiently	AgrEvo, Monsanto
• Increasing agricultural productivity while protecting nature/ minimising conflict between environmental concerns and modern agriculture/reconciling need for environmental sustainability and higher productivity/more sustainable high-yield agriculture/low-impact, high-output agriculture	AgrEvo, BASF, Bayer, Monsanto, Novartis
Feeding the world	
• Helping feed the world/providing food for developing countries	Bayer, BASF, Danisco, Monsanto, Pioneer, Rhône-Poulenc
Contributing to integrated crop management (ICM)	
• Promoting ICM as the basis of efficient and profitable production that is environmentally responsible/biotechnology can provide ICM opportunities	AgrEvo, BASF, Bayer, Zeneca
Encouraging responsible practice	
• Being responsible/encouraging environmental responsibility/ developing guidelines for responsible use of genetic engineering/using transgenic plants 'correctly'/ encouraging good environmental practice	AgrEvo, Bayer, Danisco, Rhône-Poulenc
Protecting the land for future generations	
• Protecting the sustainability of the land for future generations	Zeneca
Minimising resource use and waste	
• Saving on materials and energy/minimising waste production/using life cycle analysis/ setting environmental targets at factory level + green accounts	BASF, Bayer, Monsanto, Danisco
Improving living standards	
• Improving living standards and quality of life (through economic success and optimum use of resources)	BASF
Responding to society	
• Taking an interest in the role of companies in society	Danisco
• Encouraging open dialogue	BASF
• Include ethical and social issues in annual reports/achieving environmental and social, as well as financial sustainability: 'triple bottom line'	Danisco, Zeneca

being developed with output traits, will result in increased rather than reduced pesticide use as farmers seek to protect their investment in the costly seed. Some managers view claims about 'feeding the world' as naïve.

Several of the major biotechnology companies support integrated crop management, expressing views that correspond more closely with the 'environmental management' perspective on sustainable agriculture. They point out that GM crops will allow more precisely targeted pest control, and avoid harm to beneficial insects such as pest predators.

Some of the biotechnology companies embrace a wider view of sustainable development, beyond concerns relating specifically to agriculture. For example, they mention acting responsibly, considering future generations, minimising resource use and waste, improving living standards and responding to society's concerns and needs.

National governments such as in the US and UK, and regional institutions such as the European Commission, have strongly supported the biotechnology industry. They see it as offering employment opportunities and increasing their country's or region's competitiveness. They add the proviso that their support will not be at the expense of human health or the environment. For example, the UK's prime minister has said (Blair 2000):

> Our scientists are among the world leaders in the whole area of biotechnology. It is exactly the kind of knowledge-based industry which could help provide more jobs and more prosperity in the future. But jobs and profit will never be more important for a responsible government than concern over human health and our environment.

A number of international bodies and learned societies have published statements and reports in support of biotechnology, pointing out its potential benefits. For example, the United Nations Development Programme has emphasised biotechnology's potential to reduce malnutrition and to help poor farmers on marginal land in sub-Saharan Africa. Its report has criticised Western environmentalists for campaigns that may prevent these benefits being realised (UNDP 2001).

A report from seven learned scientific societies has listed the potential benefits from GM plants as: improved human nutrition as a result of modification of the protein, starch, fat or vitamin content of plants; plants resistant to viral, fungal and bacterial diseases; improvements to the structure and development of plants (e.g. earlier or later flowering and seed production); increased tolerance to stress such as that resulting from salinity and drought; production of extra plant biomass as a sustainable source of fuel; increased flexibility in crop management; decreased dependency on chemical insecticides; decreased soil disturbance; enhanced yields; easier harvesting, and higher proportions of the crop available for trading; decreased cost of food. Like the UN report, this report concludes that 'it is critical that the benefits of GM technology become available to developing countries' (The Royal Society of London *et al.* 2000).

Claims about potential risks

Concerns about the potential risks of GM crops have been raised not only by environmental groups, the media and the general public but also by governments and government agencies, scientists, farming organisations, food retailers and processors, consumer groups, religious groups and aid agencies. The concerns range from the narrowly scientific to much broader social ones such as the control of the technology and more intangible ones to do with uncertainties and with fundamental ethical principles.

One group of scientific concerns relates to the potential risks associated with gene flow from herbicide-tolerant GM crops to other cropped plants or related wild species, especially where these grow in close proximity. If some herbicide-tolerant GM plants persist as 'volunteers' in succeeding crops, or if herbicide tolerance is transferred to weeds, this may create weed control problems for farmers and lead to more, rather than less, use of herbicides. If GM pollen is transferred to conventional crops, or GM seed is inadvertently mixed with conventional seed during handling, this may compromise the purity or organic status of the contaminated crop or seed. Whereas some groups see these risks as soluble agronomic problems, others see any spread of GM traits as 'genetic pollution'.

In the case of Bt insect-resistant GM crops, there is concern that they may speed up the development of Bt resistance in Bt-susceptible species. This would create particular problems for organic growers, for whom sprays of Bt toxin are one of the few available pest control measures. Another concern is that insect-resistant GM crops may harm non-target species such as pollinators (e.g. bees), pest predators (e.g. ladybirds) and other valued insects (e.g. butterflies), either directly or indirectly through the food chain or through changed management practices associated with the GM crop.

The likely changes in management practices give rise to concerns about the possible risks of GM crops to farm wildlife and the ecosystem more generally. For example, there is concern that the resulting increased use of broad-spectrum herbicides may remove sources of food for birds and other species. The changed timing of herbicide use may change its environmental impact.

There are concerns about the impact of GM crops on food variety and quality. If companies invest mainly in the development of a few crops and varieties with the greatest market potential, and those are widely adopted by growers, then the production of more localised varieties and regionally typical produce may decline. Other concerns about GM foods include their implications for human health, for example that they might unexpectedly contain toxins or allergens, or that the antibiotic resistance marker genes sometimes used in the GM process might transfer to gut bacteria and make antibiotic treatment ineffective.

Among the concerns about social impacts are that GM crops will selectively benefit the larger-scale and more technologically innovative farmers, putting smaller-scale non-adopters at a disadvantage, especially in developing countries. Aid agencies are concerned that poor farmers may not be able to afford GM seed packages, or may not be able to survive if GM crops fail, for example because of the development of pest resistance. The income to developing countries from exports such as coconut oil may be threatened if temperate crops can be genetically modified to produce equivalent products.

There are concerns about institutional processes: issues to do with control of the technology, confidence in regulatory oversight of development and use, account-ability and liability. For example, there is concern about the concentration of control over GM technology in the hands of a few large multinational companies, with recent mergers among agro-chemical and seed companies serving to heighten that concern. This concentration may limit farmers' choice of crop varieties and pesticides. It may tie them to contracts that prevent them saving their own seeds. There is disquiet about the lack of transparency in company and government decisions and about the assumptions and values that underpin them. There is also concern about the influence of industry on the GM research agenda of the public sector, and about the focus on marketable GM products and patents to the possible exclusion of more systemic approaches.

At a deeper and more general level still, there are fundamental ethical concerns, for example about the rights and wrongs of 'interfering with nature'. There is public unease about confident safety predictions and assurances about GM technology when there remain large areas of uncertainty and when the unanticipated conse-quences of other innovative technologies have been all too evident.

All these concerns – environmental, social, political and ethical – have a bearing on sustainable development defined broadly. While some may be resolved by scientific evidence and further research, others that are more open-ended can only be addressed by encouraging the fullest possible discussion among well-informed people, incorpo-rating the views not only of all the stakeholders but of those who will be affected by the decisions.

Evidence of benefits

In terms of the scientific evidence, proper assessment of a novel farming technology, such as biotechnology, requires a decade's study, according to the US National Center for Food and Agricultural Policy (quoted in ENDS Report 1999). Bt cotton first became available commercially in the US in 1995, Bt corn (or maize) and herbicide-tolerant corn and soybean in 1996, and herbicide-tolerant cotton in 1997, initially in small quantities, so evidence of benefits is limited and any conclusions can only be tentative and provisional. The main evidence to date comes from annual farmer surveys in the US by the Economic Research Service (ERS) and National Agricultural Statistics Service (NASS). This section reviews that evidence, looking in turn at farmer uptake, pesticide use, yields and net returns.

Farmer uptake

One important indicator of the benefits of GM crops is their rapid rate of adoption by farmers in the US. By 1999 in the US, 57% of the total soybean area, about 55% of the cotton area and 33% of the corn area were planted with GM varieties. For all GM varieties in the US, there was an initial steep increase in adoption to 1998, followed by a much smaller increase in 1999 and signs of a levelling off or even a decrease (particularly in the case of GM corn) in 2000 (NASS 2000).

In the 1997 NASS farmer survey, farmers said their main reasons for adopting GM crops were to increase yields through improved pest control (54–76% of farmers, depending on crop and trait) and to decrease pesticide input costs (19–42%, the highest proportion being for Bt cotton). With low market prices for agricultural products at the time, it is not surprising that ways of increasing income and reducing costs were high on most farmers' list of priorities. Reasons that might be linked to reducing environmental impact were low on the list. 'Increased planting flexibility' (for example, by using reduced-tillage or no-tillage systems) was cited by only 2–6% of farmers, and 'adoption of more environmentally friendly practices' by 0–2% (Fernandez-Cornejo & McBride 2000).

The levelling off, or in some cases decline, in the rate of adoption in the US in 2000 has been put down to uncertainties about the market for GM produce. As already mentioned, in the spring of 1999, when most farmers had already made their planting decisions, some major US grain processors announced that they would not buy GM corn unless it was a variety that had been approved for import by the EU (ERS 1999a). Another factor may be a growing resistance among farmers to the tight controls and conditions imposed by companies. For example, Monsanto insists that farmers sign a 'technology agreement', imposing management conditions that include a ban on seed-saving (a traditional practice in the US). Farmers risk becoming 'squeezed' between two powerful industries, both imposing their own contracts and demands: the suppliers of crop protection packages concerning farm inputs, and the food processors and retailers concerning farm outputs (Fernandez-Cornejo & McBride 2000).

Pesticide use

From the limited data so far available, the evidence is that the adoption of GM crops in the US has led to overall reductions in pesticide use. Estimates based on ERS/NASS survey data from 1997 and 1998 for soybean, cotton and corn suggest overall reductions of between 7.6 million and 19 million acre-treatments (number of acres multiplied by number of pesticide treatments) and between 0.3 million and 8.2 million pounds of active ingredient (ERS 2000). These broad-brush figures conceal numerous complicating factors, for example the methods of analysis and the assumptions they incorporate, differences in the characteristics of adopters and non-adopters of the technology, variations by crop type and GM trait, differences in pest pressure, regional and climatic factors, and the different impacts that result from changes in the mix of pesticides used. This section first describes the different methods of analysis that have been used to show why claims about the impact of GM crops on pesticide use have to be treated with some caution. It then examines the information available so far on pesticide use for the main GM crop–trait combinations.

Method of analysis

The ERS has used three methods to estimate the changes in pesticide use due to the adoption of GM crops:

(1) differences in pesticide use between adopters and non-adopters in the same year;
(2) differences in pesticide use from year to year, based on all farmers but adjusted for the proportion of GM to non-GM crop area;
(3) differences in pesticide use from year to year, adjusted for variations due to such factors as farmer characteristics and pest pressure by the use of an econometric model.

The first method does not take into account the likelihood that adopters may differ from non-adopters in ways that affect the analysis. The ERS examined farmer variables such as farm size, farmer education and experience, debt-to-assets ratio, use of marketing or production contracts, use of irrigation and use of consultants. They found that the adoption of herbicide-tolerant soybean was linked to large farm size, better education and use of reduced-tillage or no-tillage practices (ERS 1999c). Adoption was also positively linked to average crop price and weed infestation levels.

The second method does not take into account variation in pesticide use from year to year due to factors other than the use of GM technology. For example, variation in pest incidence may affect the use of pesticides. Farmers may cut back on their use of pesticides when input costs rise or when the expected market prices for their crops are low. There may be an underlying trend towards reduced pesticide use to comply with the demands of food processors or environmental policies, for example by increased adoption of integrated crop management.

The third method attempts to take as many as possible of these 'other factors' into account by using a two-stage econometric model. The first stage examines the adoption decision for GM crops and for other pest management practices that might affect pesticide use. The second stage incorporates the output from the first and examines the impact of using GM crops on yields, net returns and pesticide use (Fernandez-Cornejo & McBride 2000).

Even after as many of the 'other factors' have been taken into account as possible, there remains the question of what the overall figures for a reduction in pesticide use mean in terms of environmental impact (*cf.* Parris & Yukoi in this volume). The adoption of GM crops alters the mix of pesticides used. The active ingredient of one pesticide may differ in its environmental impact from the equivalent amount of active ingredient of another. For example, the use of glyphosate-tolerant soybean leads to an increase in the use of glyphosate but a more substantial decrease in the use of other herbicides. Glyphosate persists in the environment only half as long as the herbicides it commonly replaces (imazethapyr, pendimethalin and trifluralin in the case of soybean) and is much less toxic to humans (ERS 2000). So in this case there is an environmental benefit both in terms of an overall reduction in pesticide use and in the switch to less persistent and less toxic pesticides. To take a different example, the use of Bt crops may reduce the use of pesticides to control those pests affected by the Bt toxin, but they still require the use of pesticides to control the pests that are unaffected. The mix of pesticides used on Bt crops may change and may even increase in years and regions where the pests unaffected by Bt cause serious problems (for example, the boll weevil in Mississippi in 1997 on cotton) (ERS 1999b). So assessment of the overall change in environmental impact requires a detailed analysis of the persistence and toxicity, as well as the quantity, of each pesticide used for each GM crop–trait combination.

More detail on pesticide use for the main GM crop–trait combinations grown in the US (insect-resistant cotton and corn, and herbicide-tolerant soybean, cotton and corn) is provided in the sections below. The regions referred to are farm-resource regions devised by the US Economic Research Service to divide the US into nine areas with similar types of farms and similar soils and climate.

Insect-resistant (Bt) cotton
The target pests for Bt cotton are the cotton bollworm (*Helicoverpa zea*), the pink bollworm (*Pectinophora gossypiella*) and the tobacco budworm (*Heliothis virescens*). Tobacco budworm has developed resistance to the pyrethroid insecticides used to control these pests in conventional crops (US House of Representatives Committee on Science 2000).

Comparison of adopters and non-adopters in the US in 1997 showed significant decreases in the insecticides used by adopters to control the target pests in two of the three surveyed regions (a mean of 0.54 compared with 1.27 acre-treatments in one region and 0.31 compared with 1.95 acre-treatments in the other; ERS 1999b, Table 3). The three surveyed regions accounted for almost 60% of the total US cotton area in 1997 (Fernandez-Cornejo & McBride 2000, Table 6). However, the decreases were accompanied by a significant increase in the insecticides used to control other pests in one of the three regions (8.19 compared with 4.43 acre-treatments; ERS 1999b, Table 3).

Econometric analysis of the 1997 ERS/NASS data, excluding the regions where non-target pests caused serious problems and focusing on southeast US, showed no significant change in the use (acre-treatments) of organophosphate and pyrethroid insecticides, but a significant decrease in other insecticides used on cotton (e.g. aldicarb, chlorpyrifos, oxamyl, endosulfan and difocol) (Fernandez-Cornejo & McBride 2000).

A detailed comparison of 293 farmers in the US states of Carolina, Georgia and Alabama in 1996 who grew both Bt and conventional cotton showed that on average they used 0.8 insecticide applications on their Bt crop, compared with 2.8 applications on their conventional cotton (Carlson *et al.* 1998, quoted in Ervin *et al.* 2000). The same survey showed that adopters of Bt cotton tended to use more insecticides on their conventional cotton than non-adopters (2.8 applications for adopters compared with 2.4 applications for non-adopters). Adopters are thus the farmers who stand to save most on insecticide costs by using insect-resistant cotton.

Insect-resistant (Bt) corn
The target pest for Bt corn is the European corn borer (*Ostrinia nubilalis*). In the US, this pest is only a serious problem in some years and some regions. For effective control, insecticides have to be sprayed before the corn borer larvae tunnel into the

corn stalk. Careful scouting is needed to monitor the crop for infestation above economically damaging thresholds and to ensure timely spraying. For all these reasons, insecticidal control is only used on 3–10% of the area of conventional corn in the US. Farmers who invest in Bt corn have to do so before they know the likelihood of European corn borer damage. In effect, they are using GM seed as a form of insurance against damage.

Bt corn was introduced at a time in the mid-1990s when there was an increased risk of European corn borer attack so that spraying of conventional crops to control the pest increased. Subsequently, in 1998 and 1999, levels of infestation in most areas were low. These confounding factors make it difficult to apportion changes in pesticide use due solely to the adoption of Bt corn.

Comparison of adopters and non-adopters of Bt corn in the US in 1997 showed a small but significant decrease in the insecticides used by adopters to control the target pest in Heartland, a farm-resource region with 75% of the US corn area and production that year (0.00 compared with 0.07 acre-treatments; ERS 1999b, Table 3).

Comparison of 1995 data (before Bt corn was introduced) with 1998 showed 7% reduction in the area treated with insecticides to control European corn borer, according to an analysis by the National Center for Food and Agricultural Policy (reported in Wolfenbarger & Phifer 2000). This analysis attributed 2% of the reduction to the introduction of a new chemical, and 2.5% to the adoption of Bt corn. It ignored the possibility that differences in European corn borer levels between 1995 and 1998 may have been a contributory factor.

Herbicide-tolerant soybean
Between 1996 and 1998, the overall rate of herbicide use (kg ha^{-1}) on soybean in the US declined by nearly 10% at the same time as adoption of herbicide-tolerant varieties increased from 7% –45% (Fernandez-Cornejo & McBride 2000).

Comparison of adopters and non-adopters in 1997 showed significant reductions in the herbicide use of adopters in three of the five surveyed regions (means of 1.04 to 2.09 acre-treatments for adopters compared with 2.14 to 2.62 acre for non-adopters; ERS 1999b, Table 3).

Econometric analysis of the 1997 ERS/NASS data showed a significant increase in the use of glyphosate, no change in the use of acetamides and triazines, and a significant decrease in the use of other herbicides on herbicide-tolerant soybean (Fernandez-Cornejo & McBride 2000, Table 9). The 'other herbicides' accounted for two-thirds of all herbicides used on soybean in 1997. The econometric analysis confirmed that there was a net decrease in herbicide use as a result of the adoption of herbicide-tolerant soybean. Econometric analysis of the 1998 data similarly showed an overall decrease in herbicide use associated with increased adoption of herbicide-tolerant soybean (an overall decrease of 1.8 million pounds active ingredient despite an increase of 5.4 million pounds in the use of glyphosate; ERS 2000).

Herbicide-tolerant cotton
Cotton production in the US relies heavily on herbicides to control weeds, often requiring two or more herbicides at planting and post-emergence herbicides later in

the season (ERS 1999c). Comparison of adopters and non-adopters of herbicide-tolerant cotton in 1997 found a 20% reduction in herbicide use in one region (Southern Seaboard; 3.69–4.76 acre treatments) but no significant reduction in another (Mississippi Portal). These two regions each accounted for about 20% of the total US cotton area in 1997 (ERS 1999b).

Econometric analysis of the same ERS/NASS data for 1997 found no significant differences overall between adopters and non-adopters of herbicide-tolerant cotton in their use of glyphosate, triazines or other herbicides (Fernandez-Cornejo & McBride 2000, Table 9).

Herbicide-tolerant corn
No significant difference was found in herbicide use between adopters and non-adopters of herbicide-tolerant corn in 1997 in Heartland, a region with 75% of the total US corn area that year (ERS 1999b, Table 3; Fernandez-Cornejo & McBride 2000, Table 6).

Yields

Changes in yields due to the adoption of GM crops might have an indirect environmental benefit if, as claimed, higher yields mean that less land than would otherwise be the case is needed to maintain or increase agricultural production.

In theory, Bt crops have the potential to increase yields by reducing yield losses due to the targeted pests, at least when those pests are present in damaging numbers and when control with chemical sprays is unsatisfactory. Herbicide-tolerant crops are intended to simplify weed control, so for these crops one might not expect such a direct link with yield, except as a result of more timely weed treatment. The extremely limited data so far available on the yields of commercial GM crops seem to support these expectations. Yield increases seem most likely to occur with Bt cotton, and to a lesser extent with Bt corn. The results for herbicide-tolerant corn, cotton and soybean are more variable, with signs that herbicide-tolerant cotton yields less than conventional cotton crops.

The evidence of the impacts of GM crops on yields and net returns in the following sections is based on the same ESR/NASS survey data from the US discussed in the previous section, analysed using the same approaches. So it is subject to similar limitations and complicating factors.

Insect-resistant (Bt) cotton
Comparison of Bt and conventional cotton in 1996 (one region only), 1997 and 1998 (three regions) showed a significant yield increase for the Bt crop in four of the seven region/year combinations (ERS 1999b, Table 2). Econometric analysis of the 1997 data (correcting for factors such as differences in farmer characteristics and pest infestation levels) confirmed a significant yield increase in the southeastern US.

Insect-resistant (Bt) corn
Comparison of Bt and conventional corn in 1996, 1997 (one region only) and 1998

(three regions) showed a significant yield increase for the Bt corn in two of the five region–year combinations.

Herbicide-tolerant soybean
Comparison of herbicide-tolerant and conventional soybean in 1996 (two regions), 1997 (five regions) and 1998 (six regions) showed a significant yield increase for the herbicide-tolerant crop in four of the thirteen region–year combinations. Econometric analysis of the 1997 survey data showed a small but significant yield increase overall for the five regions surveyed that year. Field trials have shown that some herbicide-tolerant soybeans can have a yield 'drag' of 6–11% compared with conventional high-yielding varieties (Benbrook 1999; Elmore *et al.*, in review; quoted in Ervin *et al.* 2000).

Herbicide-tolerant cotton
Comparison of herbicide-tolerant and conventional cotton showed reduced yields in three of four region–year combinations, with statistically significant reductions in one (Southern Seaboard 1997) (ERS 1999b, Table 2). However, econometric analysis of the 1997 survey data showed a significant increase in yields with the adoption of herbicide-tolerant cotton.

Herbicide-tolerant corn
Comparison of herbicide-tolerant and conventional corn showed a significant yield increase for the herbicide-tolerant corn in just one of the five region–year combinations surveyed (Heartland 1998) (ERS 1999b, Table 2).

Net returns

Changes in net returns might have an indirect environmental benefit if low incomes led to agricultural land being abandoned or if, as is sometimes claimed, high incomes allowed farmers to devote money to environmental improvements. Net returns affect sustainability broadly defined, so as to include social sustainability, by their effect on farming livelihoods.

Econometric analysis of the ERS/NASS 1997 data, using the relevant state average prices for inputs and outputs, showed a significant increase in the net returns from herbicide-tolerant cotton for the surveyed states as a whole and from Bt cotton in the southeast (Fernandez-Cornejo & McBride 2000, Table 9). There was no significant overall increase in net profits for herbicide-tolerant soybean. Results for the other GM crops were not mentioned.

These results help to explain the rapid uptake of GM cotton by US farmers, but do not explain why herbicide-tolerant soybean has been rapidly adopted too. Comparison of herbicide-tolerant and conventional soybean in 1997 showed that net profits varied from region to region. Net profits from GM soybean were significantly higher than those from conventional soybean in the Heartland land resource region, which had 70% of the total US soybean area in 1997 (Fernandez-Cornejo & McBride 2000, Tables 10 and 6). An alternative explanation for the rapid uptake of GM soybeans is that they became available just at the time when soybeans became eligible for support payments in the US (Directorate-General for Agriculture 2000).

To summarise, evidence of benefits from GM crops grown commercially is so far based on the analysis of two or three years' survey data at most. Results vary by crop–trait combination, by region and by season. Overall, results so far suggest that adoption of Bt cotton in the US can lead to reductions in the insecticides used against the target pests, although in some regions the use of other insecticides may be increased to cope with pests not controlled by the Bt toxin. The net environmental impact is unclear. For Bt corn, there may be some savings in insecticide use, although less than one-tenth of conventional corn in the US is usually sprayed to control the target pest: the European corn borer. For herbicide-tolerant soybean, the results suggest an overall decrease in the herbicides used, despite an increased use of glyphosate. This represents a decrease in environmental impact in terms of the persistence and toxicity of the changed mix of herbicides, as well as in terms of the overall quantity used. For herbicide-tolerant cotton and corn, there is little evidence so far of a significant overall reduction in the use of herbicides.

There are many factors that complicate the analysis of survey results. For example, it seems that adopters of GM crops are likely to have larger farms, be better educated and use more inputs than non-adopters. Even when more evidence is gathered so that the analyses can be more detailed, there remain many unanswered questions about what the resulting changes in farm management, including changes in pesticide use, may mean in terms of their environmental impact.

Evidence of risks

Both in the US and the EU, systematic independent research into the possible environmental impact of GM crops has been limited. In 2000, after some GM crops had already received EU approval for commercial use, large-scale farmer-managed trials were established in the UK to examine the effects on farmland wildlife of GM autumn- and spring-sown oilseed rape, forage maize (corn) and sugarbeet over a three-year period. A legal requirement to monitor commercial GM crops for their impact on wildlife was included in the revisions to Directive 90/220 that were agreed early in 2001. Previously, evidence of environmental impacts depended mainly on isolated pockets of research in public sector research institutes and universities. Company research to meet regulatory requirements focused mainly on agriculturally relevant impacts.

This section reviews the main research evidence available so far about the risks of GM crops, looking first at insect-resistant crops, then at herbicide-tolerant ones.

Insect-resistant (Bt) crops

As mentioned in the previous section, genes coding for the toxin from *Bacillus thuringiensis* (Bt) have been incorporated into maize (corn) to protect it from the European corn borer and into cotton to protect it from the cotton bollworm, pink bollworm and tobacco budworm. They have also been incorporated into potato to protect it from the Colorado beetle. Evidence of risks relates mainly to the possibility of pest resistance to the Bt toxin, and direct and indirect harm to non-target species.

Pest resistance to Bt toxins

Companies accept the possibility that the targeted pests will develop resistance. In laboratory tests, at least ten species of moths, two species of beetles and four species of flies have developed resistance to Bt toxins (Tabashnik 1994, quoted in Wolfenbarger & Phifer 2000). In the field, the diamond back moth (*Plutella xylostella*), a common and widespread pest of Brassica species, has developed resistance to sprays of Bt toxin. Whereas Bt sprays, which are used for pest control in organic farming, are used intermittently, the Bt toxin in Bt crops is present throughout the season. This increases pest exposure, so may increase the chances of resistance developing, especially if (as in some Bt maize, for example) the levels of Bt toxin expressed by the crop are not uniformly high or tail off towards the end of the growing season.

Companies initially assumed that the problem of pest resistance could be overcome by introducing genes coding for other types of Bt toxins into crop plants. This assumption was called into question by evidence that in some insect pests a single gene can confer resistance to four types of Bt toxin (Tabashnik *et al.* 1997; Fox 1997).

In response to the concerns of consumer groups and organic farmers, the US Environmental Protection Agency (EPA, responsible for the regulation of pesticidal crops) introduced a requirement for 'insect resistance management' to delay the development of pest resistance. Whereas in 1995, Bt potato was granted unconditional approval for commercial use, in 1996 Monsanto's Bt cotton was approved only on condition that farmers planted it with conventional cotton on 4% of the area, so as to provide reservoirs of Bt-susceptible pests. Subsequently, more stringent resistance management measures were imposed. In the autumn of 1996, the EPA imposed restrictions on the growing of Bt corn in cotton-growing areas of southern US to prevent additional selection pressures for Bt-resistance, since the corn earworm is also a pest of cotton (when it is known as the cotton bollworm) (Fox 1996). In 2000, the EPA increased the proportion of conventional corn that has to be planted with a GM corn crop to at least 20%, or 50% where cotton is also grown (Smith 2000; Dove 2001).

The refuge areas are intended to increase the chances that any resistant pests from the Bt crop will mate with susceptible pests from the conventional crop and produce susceptible offspring. This strategy is based on an assumption that susceptibility is the dominant trait. It also assumes that resistant and susceptible pests will mingle and will reach the reproductive stage at the same time. There is some evidence that these assumptions are not necessarily valid.

Laboratory studies showed that in the European corn borer Bt resistance may be partially dominant (Huang *et al.* 1999). When Bt resistant corn borer moths were mated with Bt susceptible ones, the offspring were much closer to the resistant parent than to the susceptible one in their response to the Bt toxin. Back-crossing the offspring with susceptible individuals did not result in any significant decline in resistance.

Laboratory experiments on the pink bollworm showed that although in this case resistance was recessive, resistant larvae on Bt cotton took six days longer to develop than susceptible larvae on conventional cotton (Liu *et al.* 1999). The researchers said that although so far Bt resistance in pink bollworm is not a problem, strategies for pest resistance management need to take into account the possibility of uneven larval development. If the slower development of resistant larvae limits the chances of cross-breeding between resistant and susceptible individuals, it could accelerate the development of resistance. If slower development reduces the numbers of resistant insects surviving over winter, it could delay the development of resistance.

Use of Bt cotton can lead to outbreaks of other pests that were previously controlled incidentally by the pesticides that the Bt toxin replaces. As a result, additional insecticidal sprays may be needed. For example, a four-year study (1996–1999) comparing 360 fields of Bt cotton with the same number of fields of conventional cotton in North Carolina found that although bollworm damage was more than halved in the Bt crops there was a four-fold increase in damage to the cotton bolls due to stink bugs (Bacheler 2000). Extension agents reported a considerable increase in cotton boll damage due to stink bugs in Bt cotton in southeastern US in 2000 (Hollis 2000). They recommended the use of organophosphates such as dicrotophos and methyl parathion if the stink bugs occurred in numbers above certain thresholds.

Harm to non-target species

There is some evidence of harm to non-target species such as butterflies. For example, in laboratory tests at Cornell University in which larvae of the Monarch butterfly (*Danaus plexippus*) were fed on milkweed leaves dusted with pollen from Bt corn, nearly half the larvae died, whereas none died when fed on pollen-free leaves (Losey *et al.* 1999). The surviving larvae on the Bt pollen-dusted leaves consumed less than those on pollen-free leaves and grew to only half their normal size. These results attracted widespread publicity because of the Monarch butterfly's strikingly beautiful appearance. They alarmed conservationists because, although the butterfly is not endangered, it is a migratory species that overwinters in Mexico and its migratory behaviour is threatened by the loss of wooded habitat. Its larvae feed exclusively on milkweed, which is commonly found growing near corn fields. Over half its summer population occurs in the corn belt of the mid-west US.

Further evidence of the impact of Bt corn pollen on the Monarch butterfly came from research at Iowa State University (Hansen-Jesse & Obrycki 2000). Researchers there fed Monarch larvae on samples of milkweed that had previously been placed within a Bt corn field and at varying distances from the edge of the field at the time the corn was shedding pollen. Within 48 hours of feeding on pollen-dusted leaves from within the Bt corn, larval mortality was 19% compared with 0% for larvae on milkweed taken from a conventional corn crop and 3% on milkweed leaves with no pollen. Larvae less than twelve hours old and those fed on milkweed taken from within ten metres of a Bt crop were the most affected.

To counter concerns raised by these results, the US Environmental Protection Agency at first asked companies to ask growers to position non-Bt corn refuges between their Bt corn and any milkweed. However, subsequently the EPA decided such buffer zones were of little use, since:

> during 2000 it became clear that milkweed in the cornfield, rather than outside it, is the preferred breeding place for Monarch, and the heavy pollen travels only a short distance, so that the amount on milkweed drops dramatically within a few metres from the cornfield so a buffer zone would be superfluous for protecting Monarch larvae and we did not continue our suggestion
> (EPA official, interview, L. Levidov 2000, personal communication)

The research on the Monarch butterfly prompted research into possible impacts of Bt crops on other butterfly species, for example the black swallowtail (*Papilio polyxenes*) whose host plants are found mainly along roadsides at the edge of corn fields. Researchers at the University of Illinois placed potted host plants with black swallowtail larvae at various distances from a crop of Bt corn at the time it was shedding pollen (Wraight *et al.* 2000). The amount of pollen deposited on the host plants was estimated from the amount deposited on greased slides placed nearby. Large numbers of the larvae died over the seven days of the study – as happens normally in the field according to the researchers – but there was no correlation between larval mortality and proximity to the Bt corn or to the amount of pollen present on the host plant leaves. In laboratory tests over a three-day period, pollen dusted onto leaf discs of the host plants failed to kill black swallowtail larvae even at the highest pollen dose tested (10 000 grains cm^{-2}).

How well any of these experiments reflect the conditions in commercial crops is unclear. Each has been criticised for methodological shortcomings. For example, the Cornell work was criticised for not reporting the levels of pollen deposited on the milkweed leaves, or taking into consideration a possible anti-feedant effect of the pollen. The value of the Illinois results was questioned, given the very high background level of larval mortality. The seven-day period of the Illinois study was criticised as an inadequate test of the likely impact in the field, given that black swallowtail larvae go through several generations in the same place each summer so are more likely to be exposed to corn pollen than the larvae of the Monarch butterfly.

Indirect impacts on non-target species
There is conflicting evidence on the possibility that the impact of Bt toxin might pass along the food chain to affect species that feed on pests targeted by Bt crops. Predators of pests may die if they depend solely or to a large extent on pests that are effectively controlled by Bt toxin. This seems likely to be the cause of a decline in numbers of a predator of Colorado beetle noted in Bt potato (Riddick *et al.* 1998). Species that depend on a Bt-targeted pest to complete a stage of their life cycle may die if their host dies before that stage is completed. For example, in laboratory studies the larvae of a parasitic wasp (*Cotesia plutellae*) died when forced to develop on the larvae of Bt-susceptible diamond back moth (*Plutella xylostella*) fed on Bt oilseed rape (Schuler *et al.* 1999). This was hardly surprising since the moth larvae all died within five days of feeding on Bt plants, whereas the larvae of the parasitic wasp take seven days to

develop. However, in the field behavioural factors may come into play. The parasitic wasps are attracted to the moth larvae by the chemicals released from the plants they are damaging. In wind tunnel studies, the researchers found that only 11% of the parasitic wasps were attracted to leaves of Bt oilseed rape on which Bt-susceptible moth larvae were feeding, while 89% were attracted to non-Bt oilseed rape leaves because those leaves suffered more feeding damage. With Bt-resistant moth larvae, there was no significant difference in feeding damage or wasp attraction between Bt and non-Bt plants. So in the field, behavioural factors might help parasitic wasps seek out and control Bt-resistant moth larvae because they cause more plant damage than Bt-susceptible moth larvae.

The possibility that Bt toxin may increase in activity when it is ingested by some species, rather than be broken down, has been raised by laboratory studies in which lacewing (*Chrysoperla carnea*) were reared on Bt-fed prey (Hilbeck *et al.* 1998; 1999). These found a significantly higher mortality (59–66%) for lacewing reared on prey fed on Bt corn than for lacewing reared on prey fed on non-Bt corn (37%). Similar results were obtained when the Bt toxin was incorporated directly into the lacewing diet (56% mortality, compared with 30% for lacewing fed on a Bt-free diet). The similarity in the results was unexpected since the level of Bt toxin in the corn leaves was less than 4 μg g^{-1} of fresh leaf whereas it was 25–100 μg g^{-1} of the artificial diet. The researchers speculated that the toxin might have been altered by biochemical processes in the prey in a way that was lethal to the lacewing predator but not to the prey.

There is little information on the possible impacts of Bt crops on soil organisms. Any impacts are likely to depend on the persistence of the toxin in the soil. Laboratory studies suggest that in neutral soil the toxicity of Bt toxin from Bt cotton and corn rapidly declines, so that by 120 days its effect on larval growth is 17–23% of its initial activity (Sims & Holden 1996; Sims & Ream 1997; quoted in Wolfenbarger & Phifer 2000). However, active toxin readily binds to soil particles, inhibiting breakdown by microbes (Stotzky 2000). High clay content and low pH increase the toxin's persistence (Stotzky 2000). Bt toxin may enter the soil via pollen or when Bt residues are ploughed in, but a laboratory study suggests it may also enter the soil by means of exudates from the roots of Bt corn (Saxena *et al.* 1999), in which case it may be present throughout the cropping season.

Herbicide-tolerant crops

Herbicide-tolerant crops may give rise to herbicide-tolerant weeds. There are several ways this might occur. Herbicide-tolerant crop plants might survive as volunteers in succeeding crops. The increased use of a restricted number of herbicides might exert a selection pressure that favours the survival of more herbicide-tolerant weed species or individuals. Gene transfer from a GM crop to a weedy relative might occur by hybridisation. There is little information on the possible ecological and economic consequences, except by analogy with introduced species that have become invasive. If herbicide-tolerant weeds became a problem for farmers, this might encourage the use of alternative herbicides, possibly ones which are less environmentally benign.

Selection pressure on weeds

Weed species that are inherently more tolerant of particular herbicides such as glyphosate may increase because of the adoption of herbicide-tolerant crops. Previously, glyphosate has been used for 15–20 years without resistant weeds becoming a problem. However, now there are reports of glyphosate tolerance in rigid ryegrass (*Lolium rigidum*), which is a pernicious weed (Wolfenbarger & Phifer 2000).

Gene transfer to weedy relatives

Since twelve of the world's thirteen most important food crops form hybrids with wild relatives (Ellstrand *et al.* 1999; Snow & Palma 1997; quoted in Wolfenbarger & Phifer 2000), gene transfer from GM crops to weeds is a possibility. The chances will increase as the area planted to GM crops increases, since the likelihood will increase that the crops will be planted close to weeds with which they can hybridise.

In Europe, gene transfer from maize to wild relatives is not an issue. However, glufosinate-tolerant oilseed rape can form fertile hybrids with wild turnip (*Brassica rapa*; Mikkelson *et al.* 1996; Snow *et al.* 1999) and wild radish (*Raphanus raphanistrum*; Chevre *et al.* 1997). Hybrids between oilseed rape and wild mustard (or charlock, *Sinapsis arvensis*) are considered unlikely (Lefol *et al.* 1996).

So far no maize volunteer weeds have been found in Europe. By contrast, oilseed rape sheds its seeds readily and can enter secondary dormancy, so it has become a widespread seedbank weed capable of over-wintering (Pekrun *et al.* 1998; Squire 1999). Studies of glufosinate-tolerant oilseed rape suggest that it is unlikely to survive in uncultivated areas (Crawley *et al.* 1993)

Gene transfer to conventional crops

Glufosinate-tolerant oilseed rape can form fertile hybrids with conventional oilseed rape (Scheffler *et al.* 1995) so there is a need to keep the crops separate for seed production purposes. A controversy arose in 2000 over contamination of conventional oilseed rape seed from North America with GM seed, some of which had not been approved for use in the EU. Contamination may have occurred in the field or during handling.

In the UK, researchers noted seed formation on male sterile plants of oilseed rape four kilometers away from any other oilseed rape crop, although it is possible that the pollen may have come from feral oilseed rape plants closer than this (Thompson *et al.* 1999). The National Pollen Unit in the UK reported finding pollen from GM oilseed rape in beehives 4.5 kilometers away from the crop (FoEE 1999), though this finding remains contentious. The Soil Association, representing organic farmers in the UK, has asked for the separation distance between GM crops and other crops, especially organic ones, to be increased to six kilometers. Currently the separation required is fifty meters from conventional crops, or 200 meters from organic crops and crops grown for seed production.

Indirect effects

There are concerns that the widespread use of broad-spectrum herbicides such as glyphosate will affect farmland birds by removing all the arable weeds and their associated insects that birds feed on. Researchers using population models showed that seeds of fat hen (*Chenopodium album*), which form the staple diet of skylarks, disappear when herbicide-tolerant sugarbeet and its associated herbicide are used (Watkinson *et al.* 2000). However, their conclusions have been challenged (Firbank & Forcella 2000).

Conversely, Monsanto-funded research at the Brooms Barn Sugar Beet Research Station has shown that weeds can be left to grow for longer in herbicide-tolerant crops, to provide habitat for beneficial insects and wildlife, then sprayed once competition with the crop becomes economically significant (Richardson 1998).

There are reports that the root zone of herbicide-tolerant oilseed rape (or rather, canola) has a less diverse bacterial community, with a different community structure, than the conventional crop (Siciliano *et al.* 1998; Siciliano & Germinda 1999). The impact of these changes is unknown, although they might affect plant decomposition rates and carbon and nitrogen levels (Wolfenbarger & Phifer 2000).

To summarise, research so far into the possible risks of GM crops is extremely patchy. There is evidence that the development of resistance to Bt toxin in pest species is likely. Resistance management that relies on refuges of susceptible pests may not work for some pest species. Resistance management strategies need to take into account not only the extent to which the gene for resistance is dominant or recessive but also the possibility of different rates of development between resistant and susceptible forms of the pest.

Outbreaks of secondary pests can occur on GM crops in the absence of the pesticides used previously. There is conflicting evidence of harm to non-target species. Laboratory studies may not reflect conditions in the field, especially in relation to insect behaviour.

Gene transfer can occur between some GM crops and conventional crops or weedy relatives. Again, it is unclear how likely this is to occur in the field under farm management conditions. Contamination of conventional crops with GM crops as a result of handling has already caused marketing problems.

There has been only limited research so far into the possible indirect effects of GM crops on ecosystems, or into farmer management practices.

Implications for sustainable development

The most striking aspect of this examination of the potential benefits and risks of crop biotechnology is the limited extent of the evidence available so far. This lack of evidence allows strong claims and counter-claims about biotechnology's potential contribution to sustainable development to co-exist.

Evidence of benefits relies mainly on the annual farmer surveys being conducted in the US by the Economic Research Service and the National Agricultural Statistics Service. The analyses published so far are based mainly on their data for one year, 1997, whereas some US policy advisors argue that reliable assessment of a new farming technology needs to be based on experience accumulated over ten years. It is difficult to separate out the impact of GM technology from the many confounding factors such as variations in weather conditions, pest incidence and farmers' management practices. The analysts admit that the assumptions they have built into their econometric analysis of the 1997 data to allow for these factors may not be valid for the large switches from conventional to GM cropping that have occurred year on year. Beyond the US, very different conditions may apply, so similar systematic and independent surveys need to be initiated in all regions where GM crops are grown. Changes in herbicide and insecticide use need to be analysed in detail for their causes and environmental impact. It cannot be automatically assumed that they are due to the adoption of GM crops alone or that reductions in pesticide use lead to a corresponding reduction in environmental impact.

Research into the risks, or safety, of GM crops is poorly funded and extremely patchy so that evidence of risks is even more sparse than that of benefits. While there is more baseline ecological data on arable systems than generally supposed (for example, seedbank records in the UK go back to 1915 and there is also substantial information on insects), there remain large gaps in the relevant ecological knowledge. Much of the research into the impact of GM crops is unco-ordinated, with as yet no agreed and established methodological approaches, so that the validity of results is frequently questioned. There is no general agreement on the agricultural management practices that should be used as the baseline for judging changes due to GM crops. More research is needed not only into the direct environmental impacts of the pesticides associated with GM crops but also into indirect impacts along the ecological food chain and the effect on this of adaptive behaviour in target and non-target species. Researchers are operating in a fluid context in which new knowledge, changing values and increasingly strict regulations lead to a need to question previous conclusions and establish further experiments. The uncertainties that can, and cannot, be resolved by research need to be clearly distinguished.

So, given that the evidence of benefits and risks is limited and that there remain many uncertainties, what can we conclude about the potential contribution of biotechnology (in particular, GM crops) to sustainable agriculture and sustainable development?

To consider first the 'market-based' concept of agricultural development, it is too early to judge the claims that GM crops will increase yields so that less land will be needed for agriculture than would otherwise be the case, or that they will provide farmers with extra income that will be invested in environmental improvements. For the GM crops currently in commercial production, the evidence for increased yields and income is variable. Low market prices for agricultural products have led farmers in northern America to adopt GM crops in the hope of reducing costs and maintaining viability, reasons that would be unlikely to lead to a reduction in the cropped area or to investment in environmental measures.

Nor does the adoption of GM crops appear to be compatible with the 'community-based' concept of sustainable agriculture, except insofar as it has stimulated the market for organic produce. Even then, new entrants to organic agriculture in response to the increased market demand may not share the ideals of community-based sustainable agriculture commonly associated with supporters of organic methods (*cf.* Barling in this volume). Instead, these new entrants may apply industrial-scale approaches to organic production. As far as GM crops are concerned, it could be argued that the evidence from the US that adopters of GM crops tend to be larger farmers and more intensive users of insecticide suggests that GM crops may serve to widen the gap between large-scale farmers, producing for remote markets, and smaller-scale community-based farmers.

However, it is possible that GM crops could contribute to the 'environmental management' form of sustainable agriculture. For this to occur, it will no longer be possible to view GM crops as the easy management option, as promoted by some companies. Instead, their use will require skilful management to achieve precisely targeted pest control and to avoid any secondary or long-term undesirable impacts. Growers will require expert advice, on-going training and well-funded research support. They will need to be given clear management guidelines, backed up by regulation and monitoring.

Biotechnology can also contribute 'behind the scenes' to the environmental management form of sustainable agriculture, provided companies receive appropriate and consistent policy signals. For example, gene sequencing and gene markers can speed up the identification of desirable traits for incorporation by conventional breeding. Gene technology allows rapid toxicological screening of chemicals, helping to contain the costs of identifying less harmful pesticides.

Biotechnology has the potential to contribute to sustainable agriculture, but for that potential to be realised its developers and users will need strong policy guidance. There will need to be a co-ordinated and systemic policy framework in which sustainable development is a primary and unambiguously defined aim. Other related policies, such as those on research support, pesticide use, GM regulation and agri-environmental initiatives will need to be consistent with that aim.

The definition of sustainable development that informs GM research and development will need to focus on progress in terms of social well-being rather than simply economic growth. This will involve paying more attention to the possible social impacts of GM crops (for example, increased dependence of farmers on external expertise). It will involve giving full consideration to all stakeholders' views. It will involve increasing the support for public research to encourage the development of products and processes with wider social benefits, and to allow systematic research and monitoring of potential environmental risks.

Public disquiet about GM crops and food, and about agricultural production methods more generally, has emphasised the need for a new vision to inform agricultural policy in the European Union. As mentioned at the beginning of this chapter, biotechnology presents a new opportunity to learn from the lessons of the 'green revolution'. If the new vision is to include biotechnology, then it needs not only to be 'doubly green' (Conway 1997), benefiting both agriculture and the environment. It needs to be based on sustainable development in the fullest sense, benefiting agriculture, the environment and society.

References

Bacheler, J.S. (2000) 2000 Bollgard cotton performance expectations for North Carolina producers. *Carolina Cotton Notes CCN-00-3D*, 23 March. [http://www.cropsci.ncsu.edu/ccn/2000/ccn-00-3d.htm, 25/06/2001]

Benbrook, C. (1999) Evidence of the magnitude and consequences of the Roundup Ready soybean yield drag from university-based trials in 1998. *Ag BioTech InfoNet Technical Paper No. 1.* [http://www.biotech-info.net/RR_yield_drag_98.pdf, 15/10/2000]

Blair, T. (2000) The key to GM is its potential, both for harm and good; biotechnology could bring enormous benefits to humankind, but we must proceed with caution. *The Independent (London),* 27 February, p. 28.

Carlson, G.A., Marra, M.C. & Hubbell, B. (1998) Yield, insecticide use, and profit changes from adoption of Bt cotton in the Southeast. *Beltwide Cotton Conference Proceedings,* 2, 973–4.

Carr, S. (2000) *EU Safety Regulation of Genetically-modified Crops.* The Open University, Milton Keynes.

Chevre, A-M., Eber, F. & Renard, M. (1997) Gene flow from transgenic crops. *Nature,* 389, 924.

Conway, G. (1997) *The Doubly Green Revolution. Food for All in the Twenty-first Century.* Penguin, London.

Crawley, M.J., Hails, R.S., Rees, M., Kohn, D. & Buxton, J. (1993) Ecology of transgenic oilseed rape in natural habitats. *Nature,* 363, 620–23.

DETR [Department of the Environment, Transport and the Regions] (1998) *Opportunities for Change.* HMSO, London.

Directorate-General for Agriculture (2000) *Economic Impacts of Genetically Modified Crops on the Agri-Food Sector: A Synthesis.* (Working Document.) Directorate-General for Agriculture, Brussels.

Dove, A. (2001) Survey raises concerns about Bt resistance management. *Nature Biotechnology,* 19, 293–4.

Ellstrand, N.C., Prentice, H.C. & Hancock, J.F. (1999) Gene flow and introgression from domesticated plants into their wild relatives. *Annual Review of Ecological Systems,* 30, 359.

Elmore, R.W., Roeth, R.W., Nelson, L.A. *et al.* (in review) Yield penalties of glyphosate-tolerant soybean cultivars relative to sister lines. *Agronomy Journal.*

ENDS Report (1999) US studies call benefits of GM crops into question. *ENDS Report,* 294, 10–11.

ENDS Report (2000) GM crop businesses face new crisis over food safety and segregation. *ENDs Report,* 311, 11–12.

ENDS Report (2001) Moratorium on new GMOs set to stay despite new Directive. *ENDS Report,* 313, 49–50.

ERS [United States Economic Research Service] (1999a) Update on Bt corn and other new technology. *In Feed Yearbook/FDS-1999/April 1999,* 9-10. ERS, USDA, Washington, DC.

ERS [United States Economic Research Service] (1999b) Genetically engineered crops for pest management. ERS, USDA. [http://www.econ.ag.gov/whatsnew/issues/biotech/, 30/06/1999]

ERS [United States Economic Research Service] (1999c) Impacts of adopting genetically engineered crops in the US: preliminary results [http://www.econ.ag.gov/whatsnew/issues/gmo, 21/06/2000]

ERS [United States Economic Research Service] (2000) Genetically engineered crops: has adoption reduced pesticide use? *Agricultural Outlook*, August 2000, ERS-AO-273. ERS, USDA. [http://usda.mannlib.cornell.edu/reports/erssor/economics/ao-bb/2000/ao273f.asc, 21/08/2000]

Ervin, D.E., Batie, S.S., Welsh, R., Carpentier, C.L., Richman, N.J. & Schulz, M.A. (2000) *Transgenic crops: An environmental assessment.* Winrock International, Morrilton AR, [http:www.winrock.org,11/06/2001]

Fernadez-Cornejo, J. & McBride, W.D. (2000) *Genetically engineered crops for pest management in US agriculture: Farm-level effects.* Economic Research Service, USDA, Washington, DC. [www.ers.usda.gov/epubs/pdf/aer786/]]

Firbank, L.G. & Forcella, F. (2000) Genetically modified crops and farmland biodiversity. *Science*, 289, 1481–2.

FoEE (1999) Genetically modified crops: Genetic pollution proved. *FoEE Biotech Mailout*, 5 (7), 2. [www.foeeurope.org/biotechnology/about.htm]

FoEE (2001a) Directive 90/220/EEC: The final countdown. *FoEE Biotech Mailout*, 7 (1), 5. [www.foeeurope.org/biotechnology/about.htm]

FoEE (2001b) Traceability of GMOs. *FoEE Biotech Mailout*, 7 (2), 4. [www.foeeurope.org/biotechnology/about.htm]

Forum for the Future (1998) *Opportunities for change: A Response by Forum for the Future to the Consultation Paper on a Revised UK Strategy for Sustainable Development.* Forum for the Future, London.

Fox, J. (1996) Bt cotton infestations renew resistance concerns. *Nature Biotechnology*, 14, 1070.

Fox, J. (1997) EPA seeks refuge from Bt resistance. *Nature Biotechnology*, 15, 409.

GeneWatch UK (1999) *Genetic engineering: A review of developments in 1998.* GeneWatch, Buxton.

GeneWatch UK (2000a) *GM crops and food: A review of developments in 1999.* GeneWatch, Buxton.

GeneWatch UK (2000b) *Genetically modified crops in the UK: The current situation.* Briefing 1 September 2000. GeneWatch, Buxton.

GeneWatch UK (2001) *Genetic engineering: A review of developments in 2000.* GeneWatch, Buxton.

Hails, R.S., Ree, M., Kohn, D.D. & Crawley, M.J. (1997) Burial and seed survival in *Brassica napus* subsp. *oleifera* and *Sinapis arvensis* including comparison of transgenic and non-transgenic lines of the crop. *Proceedings of the Royal Society of London*, B 264, 1–7.

Hansen-Jesse, L.C. & Obrycki, J.J. (2000) Field deposition of Bt transgenic corn pollen: Lethal effects on the Monarch butterfly. *Oecologia*, (19 August) [http://link.springer.de/link/service/journals/00442/contents/00/00502, 15/10/2000]

Hilbeck, A., Baumgartner, M., Padruot, M.F. & Bigler, F. (1998) Effects of transgenic *Bacillus thuringiensis* corn-fed prey on mortality and development time of immature *Chrysoperla carnea* (Neuroptera: Chrysopidae). *Environmental Entomology*, 27 (2), 480–7.

Hilbeck, A., Moar, W.T., Pusztai-Carey, M., Filippini, A. & Bigler, F. (1999) Prey-mediated effects of Cry1Ab toxin and protoxin and Cry2A protoxin on the predator *Chrysoperla carnea. Entomoligia Experimentalis et Applicata*, 91, 305–316.

Hollis, P.L. (2000) Stink bugs emerging as major pest in southeast cotton. *Farmsource*, 27 April. [http://www.farmsource.com, 15/05/2000]

Huang, F., Buschman, L.L., Higgins, R.A. & McGaughey, W.H. (1999) Inheritance of resistance to *Bacillus thuringiensis* toxin (Dipel ES) in the European corn borer. *Science*, 284, 965–7.

James, C. (2000) Global status of commercialized transgenic crops: 2000. ISAAA Briefs No. 21, 2000 preview. ISAAAA, Ithaca, NY. [http://www.isaaa.org, 11/07/2001]

Lefol, E., Danielou, V. & Darmency, H. (1996) Predicting hybridization between transgenic oilseed rape and wild mustard. *Field Crops Research*, 45, 153.

Levidov, L. (2000) Which sustainability? Policy dilemmas over GM crops. Paper presented to the third POSTI International Conference, Docklands Campus, University of East London, London, 1–3 December 2000. [http://www.esst.uio.no/posti/workshops/levidov.html, 11/10/2001]

Levidov, L., Carr, S. & Wield, D. (2000) Genetically modified crops in the European Union: regulatory conflicts as precautionary opportunities. *Journal of Risk Research*, 3 (3), 189–208.

Liu, Y-B., Tabashnik, B.E., Dennehy, T.J., Patin, A.L., Bartlett, A.C. (1999) Development time and resistance to Bt crops. *Nature*, 400, 519.

Losey, J., Rayor, L., Carter, M. (1999) Transgenic pollen harms monarch larvae. *Nature*, 399, 214.

Mikkelson, T.R., Andersen, B. & Jorgensen, R.B. (1996) The risk of crop transgene spread. *Nature*, 380, 31.

NASS [United States National Agricultural Statistics Service] (2000) Farmer Reported Biotechnology Varieties [http://usda.mannlib.cornell.edu/reports/nassr/field/pcp-bbp/psp10300.txt, 21/06/2000].

Pekrun, C., Hewitt, J.D.J. & Lutman, P.J.W. (1998) Cultural control of volunteer oilseed rape. *Journal of Agricultural Science*, 130, 155–63.

Richardson, D. (1998) A close inspection of a trial site for GM sugarbeet added weight to the arguments in favour of genetic modification. *Farmers Weekly*, 4 September.

Riddick, E.W., Dively, G. & Barbosa, P. (1998) Effect of a seed-mix deployment of Cry3A-transgenic and non-transgenic potato on the abundance of *Lebia grandis* (Coleoptera: Carabidae) and *Coleomegilla maculata* (Coleoptera: Coccinellidae). *Annals of the Entomological Society of America*, 91, 647–53.

Royal Society of London, USA National Academy of Sciences, Brazilian Academy of Sciences, Chinese Academy of Sciences, Indian National Academy of Sciences and Third World Academy of Sciences (2000) *Transgenic Plants and World Agriculture*. Royal Society: London.

Saxena, D., Flores, S. & Stotzky, G. (1999) Insecticidal toxin in root exudates from Bt corn. *Nature*, 402, 480.

Scheffler, J.A., Parkinson, R. & Dale, P.J. (1995) Evaluating the effectiveness of isolation distances for field plots of oilseed rape (*Brassica napus*) using a herbicide-resistance transgene as a selectable marker. *Plant Breeding*, 114, 317–21.

Schuler, T.H., Potting, P.J., Denholm, I. & Poppy, G.M. (1999) Parasitoid behaviour and Bt plants. *Nature*, 400, 825.

Siciliano, S.D. & Germida, J.J. (1999) Taxonomic diversity of bacteria associated with the roots of field-grown transgenic *Brassica napus* cv. Quest, compared to the non-transgenic *B. napus* cv. Excel and *B. rapa* cv. Parkland. *FEMS Microbiology Ecology*, 29 (3), 263–72.

Siciliano, S.D., Theoret, C.M., de Freitas, J.R., Hucl, P.J. & Germida, J.J. (1998) Differences in the microbial communities associated with the roots of different cultivars of canola and wheat. *Canadian Journal of Microbiology*, 44, 844–51.

Sims, S.R. & Holden, L.R. (1996) Insect bioassay for determining soil degradation of *Bacillus thuringiensis* subsp kurstaki CryIA(b) protein in corn tissue. *Environmental Entomology*, 25 (3), 659–64.

Sims, S.R. & Ream, J.E. (1997) Soil inactivation of the *Bacillus thuringiensis* subsp. kurstaki CryIIA insecticidal protein within transgenic cotton tissue: Laboratory microcosm and field studies. *Journal of Agricultural and Food Chemistry*, 45, 1502–1505.

Smith, N. (2000) *Seeds of Opportunity: An assessment of the benefits, safety and oversight of plant genomics and agricultural biotechnology (Committee Print 106–B)*. Committee on Science, US Congress, Washington, DC.

Snow, A.A. & Palma, P.M. (1997) Commercialization of transgenic crops: Potential ecological risks. *Bioscience*, 47, 86–96.

Snow, A.A., Andersen, B. & Jorgensen, R.B. (1999) Costs of transgenic herbicide resistance introgressed from *Brassica napus* into weedy *B. rapa*. *Molecular Ecology*, 8 (4), 605–615.

Squire, G.R. (1999) Temperature and heterogeneity of emergence time in oilseed rape. *Annals of Applied Biology*, 135, 439–47.

Stotzky, G. (2000) Persistence and biological activity in soil of insecticidal proteins from *Bacillus thuringiensis* and of bacterial DNA bound on clays and humic acids. *Journal of Environmental Quality*, 29, 691–705.

Tabashnik, B.E. (1994) Evolution of resistance to *Bacillus thuringiensis*. *Annual Review of Entomology*, 39, 47.

Tabashnik, B.E., Liu, Y-B., Finson, N., Masson, L. & Heckel, D.G. (1997) One gene in diamondback moth confers resistance to four *Bacillus thuringiensis* toxins. *Proceedings of the National Academy of Sciences*, 94, 1640–44.

Tait, J, Chataway, J. & Wield, D. (2001) *Policy Influences on Technology for Agriculture (PITA): Chemicals, Biotechnology and Seeds*. Final Report and Annexes C2-C16, DG Research, Project No. PL 97/1280, European Commission, Brussels. [See http://technology.open.ac.uk/cts/pita/ for the reports of the PITA project]

Tait, J. & Morris, R. (2000) Sustainable development of agricultural systems: Competing objectives and critical limits. *Futures*, 32, 247–60.

Thompson, C., Squire, G.R., Mackay, G, Bradshaw, J, Crawford, J. & Ramsay, G. (1999) Regional patterns of geneflow and its consequences for GM oilseed rape. In: *Geneflow in Agriculture: Relevance for Transgenic Crops*. BCPC Symposium Proceeding, 72, 95–100.

UNDP [United Nations Development Program] (2001) *Human Development Report 2001: Making New Technologies Work for Human Development*. UNDP, New York.

US House of Representatives, Committee on Science, Sub-committee on Basic Research (2000) *Seeds of Opportunity: An assessment of the benefits, safety and oversight of plant genomics and agricultural biotechnology* (Committee Print 106-B). Government Printing Office, Washington, DC. [http://www.house.gov/science/smithreport041300.pdf]

Watkinson, A.R., Freckleton, R.P., Robinson, R.A. & Sutherland, W.J. (2000) Predictions of biodiversity response to genetically modified herbicide-tolerant crops. *Science*, 289, 1554–7.

Wraight, C.L., Zangeri, A.R., Carroll, M.H. & Berenbaum, M.R. (2000) Absence of toxicity to *Bacillus thuringiensis* pollen to black swallowtails under field conditions. *Proceedings of the National Academy of Science*, 97 (14): 7700–7703.

Wolfenbarger, L.L. & Phifer, P.R. (2000) The ecological risks and benefits of genetically engineered plants. *Science*, 290, 2088–92.

Chapter 12

European Food Industry Initiatives Reducing Pesticide Use

Nicolien M. van der Grijp

Introduction

For many years, the reduction of pesticide use by farmers has been dealt with exclusively by public policymakers, farmers' organisations and environmental NGOs. Their policies, however, have not always been as effective as was hoped for. More recently, the influencing of farmers has taken a new turn, because companies in the food industry and retail trade have started to take initiatives aimed to increase the market share of more sustainably produced food products. These market driven initiatives fall roughly into two categories, on the one side the development of programmes to define and implement integrated production techniques or Integrated Crop Management – the option of 'fewer chemical inputs' – and on the other side activities to promote organic production and consumption – the option of 'no chemical inputs'.

At present, organic agriculture is the only type of the more sustainable agricultural production methods with an internationally recognised certification system for its products, which makes them identifiable in the market and suitable for the payment of premium prices. Products produced under protocols of integrated farming are more difficult to recognise in the market, as companies seem to have more difficulties in communicating the 'fewer chemical inputs' message than the 'no chemical inputs' message. However, the European retailing industry is presently taking steps to develop a business-to-business system to control and certify Integrated Crop Management (ICM) products and their supply chains on a European-wide scale.

The chapter will go into more detail about the nature of the initiatives being developed by the food industry, illustrated with practical examples. With this in mind, the emphasis will be on the initiatives developed by the large European supermarket chains and their umbrella organisations, because corporate retailers have gained significant market share and market power in the last decade, and are expected to strengthen their dominant market position ever more. Consequently, they wield huge power over production and consumption processes.

The chapter will partly draw on two research projects conducted by the Institute for Environmental Studies (IVM). The first one was an inventory of food industry initiatives aimed at more sustainable agricultural practices in thirteen member states of the EU plus Norway and Switzerland. It resulted in the report 'Green supply chain initiatives in the European food and retailing industry' (van der Grijp & den Hond 1999). The second research project was in fact a follow-up to the first project, and aimed to provide more in-depth insights into the situation in the Dutch food and retailing industry. It focused especially on the bottlenecks encountered by food companies and the interactions between private initiatives and public policy. This project ended in early 2001 with the publication of a report in Dutch (van der Grijp *et al.* 2001).

The structure of the chapter is as follows. The next section elaborates on the shifting power in food supply chains, and the increasing emphasis that is put by retailers on food quality and safety. The third section deals with the issue of standard setting and certification for integrated as well as organic production. The chapter continues with an overview of the current market situation for sustainably produced food. The next section goes into more detail about specific retailer initiatives in four European countries, including Italy, The Netherlands, Switzerland and the United Kingdom. The sixth section deals with the bottlenecks and challenges surrounding food industry initiatives, and the last section sketches the future perspectives for market-driven initiatives.

Shifting power in food supply chains

Supply chains of food include all human-organised activities from agriculture through food processing and retailing to the food service sector and households (Green 2000). Traditionally, suppliers, especially food processors and providers of agricultural inputs, have been the most powerful organisations in the chains. In the last decade, however, distributors, including wholesalers, retailers and restaurants, have grown in importance as shapers of both supply and demand.

One of the clear signals of the shifting power in food supply chains is the ever-increasing market share in general food sales of the major retailers in Europe and the USA, at the expense of independent supermarkets and specialist shops. On top of that, the retailing sector is subjected to ongoing processes of concentration. In several EU member states, most notably, Denmark, The Netherlands, France and the UK, the three largest retail chains account for 40–60% of the grocery market (Marsden *et al.* 2000). This means that a relatively small group of retailers can exercise a large influence over the other stakeholders in the supply chain, and is able to influence the food choices on offer. Consequently, their practices and policies are of critical importance in steering farmer supply and consumer demand, for example for products produced under organic or integrated production protocols. As Browne *et al.* (2000) put it, retailers may be regarded as 'the agents of change' in the process towards sustainability.

Retailers, and other companies in the food industry, exercise their influence by actively managing their supply chains. It is partly through their supply chains that they are seeking to secure competitive advantage. This may imply that the food industry is creating its own standards which go beyond those required by traditional forms of public regulation (Marsden *et al.* 2000). Moreover, several events in the

1990s show that supply chain management is getting increasingly institutionalised by the development of private-interest regulatory systems.

This trend of active supply chain management in combination with standard setting goes hand in hand with an increased attention to other product characteristics than price, such as quality, safety, and environmental and ethical performance. The content of these concepts is not a given, but they are constructed over and over again by retailers and food producers. It should be noted that for some market players all these aspects fall under the heading of quality and safety, and that others consider environmental and ethical characteristics as extra product qualifications. The Dutch retailer Albert Heijn, for example, has developed the concept of the pyramid with five layers to demonstrate the priorities within supply chain management (Hertzberger 1999). Each product has to meet the criteria of the bottom three layers consisting of availability, product safety, and product quality. The top two layers of the pyramid – environment and ethics – should be applied to a selection of products.

Apart from the question of whether environmental issues belong to the quality and safety domain or are something separate, the conclusion is that companies in the food industry are increasingly paying attention to environmental issues, including agricultural production methods, and, as is especially relevant in this volume's context, the use of pesticides.

Role of standard setting, certification and labelling

One of the tools used to achieve more sustainable systems of agriculture is the labelling and certification of products, services and production methods that meet certain environmental standards. This tool is certainly relevant in the context of market-driven initiatives, because its application makes the products produced under certain protocols more easily identifiable for other stakeholders in the supply chain. To stimulate participation of farmers and to strengthen consumer confidence, it is of the utmost importance to devise transparent systems of standard setting and certification against reasonable costs. Among the first requirements of transparency are a clear definition of the alternative agricultural production method, operationalisation in unambiguous guidelines and efficient procedures in principle open to all interested parties.

The next two subsections deal with standard setting and certification for organic production and production under ICM respectively. It is a noticeable fact that private parties have been, and are still, the driving forces behind the development of both systems: the grassroots organisation IFOAM is playing a central role in organic standard setting and certification, while the large European retailers are providing the main impetus for developing a system for integrated production.

Standard setting and certification for organic production

IFOAM, the International Federation of Organic Agricultural Movements, is the international umbrella organisation of organic agriculture. It was founded in 1972 by five national organic grassroots organisations. It now has 770 member organisations in 107 countries. Since its inception, it has fulfilled an important role in the development of organic standard setting. The federation also represents the organic movement in international parliamentary, administrative and policymaking forums. It has, for example, consultative status with the UN and FAO.

IFOAM defines organic agriculture as the following (http://www.ifoam.org, 21 September 2000):

> Organic agriculture includes all agricultural systems that promote the environmentally, socially and economically sound production of food and fibres. These systems take local soil fertility as a key to successful production. By respecting the natural capacity of plants, animals and the landscape, it aims to optimise quality in all aspects of agriculture and the environment. Organic agriculture reduces external inputs by refraining from the use of chemical-synthetic fertilisers, pesticides and pharmaceuticals. The use of genetically modified organisms is excluded.

IFOAM is responsible for the development of Basic Standards for Organic Production and Processing. These standards cannot be used for certification on their own. They provide a framework for certification programmes world-wide to develop their own national or regional standards. So far, at least one hundred sets of different standards have been developed at the national and regional level. These standards are very varied, due to social, cultural, economic and geo-climatic conditions, but they all require regular inspections of producers and manufacturers, and certification according to strict standards. The certification process focuses on the methods and materials used in production.

The IFOAM Basic Standards are revised according to a timetable. Within two years after a revision, all national and regional certification programmes must have incorporated the changes in their own standards. In 1992, IFOAM started with an accreditation programme to ensure equivalency of certification programmes world-wide. Since 1998, the programme has been accompanied by a consumer logo, that should provide visible reassurance to consumers in countries other than those where the product originated.

The EU devised legislation defining organic agriculture in 1991.[1] This legislation is largely based on the IFOAM system of inspection and certification. In the preamble to the Regulation the Council considers that (2092/91/EC):

> A framework of Community rules on production, labelling and inspection will enable organic farming to be protected in so far as it will ensure conditions of fair competition between the producers of products bearing such indications and give the market for organic products a more distinctive profile by ensuring transparency at all stages of production and processing thereby improving the credibility of such products in the eyes of consumers.

The EU legislation applies to unprocessed agricultural products from vegetable and, since more recently, animal origin, as well as processed food products composed of one or several ingredients. Annex 1 to the Regulation sums up the minimum requirements for organic production at farm level. According to the Regulation, a two-year transition period to convert from conventional to organic production is required for certification. This is an important fact because only certified organic products can command premium prices in the market.

The Regulation entered into force on 1 January 1993. Following its requirements, all EU member states have by now designated certification bodies which all have developed their own organic standards, inspection schemes and logos. The EU legislation also opens the EU organic market for products from non-EU countries. Their access is based on the concept of equivalence; that is, production, processing, documentation, and inspection must all be equivalent in the exporting country. Imports are allowed from countries explicitly registered, or on a case-by-case basis which involves an import authorisation procedure. However, market access for countries outside the EU proves to be rather cumbersome in practice, and even within the EU market access is not always easy, because of present tendencies to strengthen local and regional organic markets. To give a further incentive to the development of the organic market, the European Council introduced a Community logo in 2000.[2] This logo may be used alone or in combination with other national or private logos used to identify organic products. Table 12.1 summarises the development of standard setting and certification for organic production.

Table 12.1   Development of standard setting and certification for organic production.

Year	Event
1920–1930s	Introduction of organic agriculture
1972	Foundation of IFOAM
1975	First IFOAM Basic Standards for Organic Production and Processing
1992	IFOAM introduces its accreditation programme for certification bodies
1993	Entry into force of EC Regulation 2092/91 defining organic agriculture
1998	IFOAM introduces a world-wide consumer logo
1999	First Codex Alimentarius Guidelines for the Production, Processing, Labelling and Marketing of Organically-produced Food
2000	European Commission introduces a Community-wide consumer logo

Standard setting and certification for integrated production

Integrated production or Integrated Crop Management (ICM), though much used and debated, is not easily defined. This production method could be considered as a step-wise implementation of a range of agricultural practices that more or less radically diverge from conventional agriculture. ICM aims to minimise the use of fertilisers and pesticide products by favouring other measures such as natural predators, crop rotation and mechanical weeding. Pests need not be eliminated, but rather kept under control, at levels below which they cause economic damage. In this manner it encompasses the earlier concept of 'integrated pest management' (IPM) (Perkins 1982).

In a study which gives an overview of ICM practices in Europe, it was concluded that most programmes have the following common elements (EUREP 1998):

- application of only registered pesticides
- preference for non-chemical measures
- optimisation of pesticide use
- strict documentation
- regular controls.

Supporters of ICM see it as a 'quiet revolution', winning the best of both worlds by marrying organic techniques with the option of chemicals if things go wrong; others consider it a half-way house, and sceptics argue that without (legally) binding rules, ICM can mean more or less anything (see Browne *et al.* 1999; Morris & Winter 1999).

Switzerland is the country of origin of ICM, as in 1978 the SAIO working group was established which aimed at the development of standards for integrated production of fruit (EUREP 1998). Several countries followed the Swiss example, but it was only in the late 1980s that production under ICM became a serious undertaking. In that period several ambitious programmes were set up and accompanying certification schemes were developed. Among the most remarkable examples in the field are the programmes developed in Switzerland, the UK, Italy and The Netherlands. Table 12.2 provides an overview of these programmes and a few of their characteristics. Some of them will be dealt with more extensively later in the chapter, where specific initiatives in the respective countries are highlighted.

Table 12.2   ICM-initiatives in four European countries (based on Eurep 1998).

Country	Initiative	Initiator	Starting year
Italy	Almaverde	Large growers' organisation Apofruit	1987
	Prodotti con Amore	Retailer Coop	1988
	Progetto Qualità	Retailer Conad	n.a.
	Naturama	Retailer Esselunga	n.a.
	Regional trademarks	Regional growers' organisations	Varies
Netherlands	Milieubewuste Teelt	Growers' organisation	1990
	Aarde & Waarde	Retailer Albert Heijn	1991
	AgroMilieukeur	Semi-governmental organisation	1993
Switzerland	SAIO- guidelines and minimum standards for the production of fruit	Schweizerischen Arbeitsgemeinschaft für integrierte Obstproduction (SAIO)	1978
	SOV labelling scheme	Schweizerischen Obstverband (SOV)	1990
	Ökoplan	Retailer Coop	n.a.
	M-Sano	Retailer Migros	n.a.
UK	Assured Produce	Partnership of National Farmers' Union and seven retailers	1991

In 1997, a group of thirteen large European retailers founded the Euro-Retailer Produce Working Group (EUREP), with the aim of making a first step towards European-wide harmonisation of minimum standards for production under ICM (EUREP 1998). It introduced the EUREPGAP protocol in 2001, which contains the basic requirements for Good Agricultural Practice (GAP) for fruits and vegetables. EUREP expects to develop additional protocols for flowers and ornamentals, livestock, combinable crops and feed.

The membership of EUREP consists of three groups, including retail members, supplier members and associate members. Among the supplier members are growers and growers organisations from all continents. The group of associate members is of a varied composition, including certification bodies, consulting firms and the crop protection industry. As is shown in Table 12.3, the retail membership of EUREP has quickly expanded and now consists of 22 retailers from ten European countries (http://www.eurep.org, 9 October 2001). In the course of the years 2000 and 2001, Asda (UK), Coop Norway (No), Eroski (Sp), Marks and Spencer (UK), Somerfield (UK), Superquinn (Ire), Superunie (NL), and Trade Service Netherlands BV (NL) became new members. The two French participants (Continent and Promodès), though, discontinued their membership in 2001, and some other countries, for example Germany, Denmark and Switzerland, are not involved at all. Taken together, the 22 retailers are a significant market player in the European market for fruits, vegetables and potatoes, as their current sales amount to more than 25% of the European total (http://www.nak.nl, 6 September 2000).

Table 12.3 Membership of EUREP, October 2001 (http://www.eurep.org, 9 October 2001).

1. Ahold (NL)	12. Laurus (NL)
2. Albert Heijn (NL)	13. Marks and Spencer (UK)
3. Asda (UK)	14. Safeway (UK)
4. COOP Italia (It)	15. Sainsbury's (UK)
5. COOP Norway (No)	16. Somerfield (UK)
6. Delhaize "Le Lion" (Be)	17. Spar Österreich (Au)
7. DRC / Belgium Austion Market (Be)	18. Superquinn (Ire)
8. Eroski (Sp)	19. Superunie (NL)
9. ICA (Sw)	20. Tesco (UK)
10. Kesko (Fi)	21. Trade Service Netherlands BV (NL)
11. Kooperativa Förderbund (KF) (Sw)	22. Waitrose (UK)

The UK is especially strongly represented in EUREP, since four of its largest supermarket chains were original members and three others joined later. In addition, EUREP's chairman is provided by the British retailer Safeway. So it may not be surprising that the basic idea of establishing EUREP has come from the UK. British retailers participating in the Assured Produce Scheme in the UK have taken the lead in the EUREP initiative because they aimed to impose similar standards on overseas suppliers as they already did on national suppliers (EUREP 1998). Their interest in doing so is strongly linked to the entry into force of the Food Standards Act of 1990 which placed an increased liability on British retailers and food producers for the activities of other participants in food supply chains.

The present success of the EUREP initiative, including outside the UK, may be due (according to the organisation itself) to the fact that retailers are resourcing globally and are facing increasing competition, pressure on profitability and an ever-tightening regulatory environment (http://www.eurep.org, 9 October 2001). In addition food safety has become a top priority for many retailers. EUREP, therefore, estimates that its prospects for growth are 'quite outstanding'. It has the ambition to become the global player in agricultural production standards and verification frameworks for fruits and vegetables.

As already mentioned, the focus of EUREP activities is the EUREPGAP protocol for fresh fruits and vegetables. The protocol should be regarded as a basic standard. Each participating retailer can additionally use individual standards, which might be more stringent (EUREP 1998). The GAP protocol is seen as a benchmark to assess current practice, and provide guidance for further development. It will be subject to future amendments.

The protocol itself contains two lists: one of required measures (the 'musts') and one of encouraged measures (the 'shoulds') (http://www.eurep.org, 9 October 2001). Concerning pesticide use, the 2001 version of the protocol includes the following items: basic elements of crop protection; choice of chemicals; advice on quantity and type of pesticide; records of application; safety, training and instructions; protective clothing and/or equipment; pre-harvest interval; spray equipment; disposal of surplus spray mix; pesticide residue analysis; pesticide storage; empty pesticide containers; and obsolete pesticides. The requirements regarding four of these items are here further explained:

(1)   Basic elements of crop protection: 'Protection of crops against pests, diseases and weeds must be achieved with the appropriate minimum pesticide input. Wherever possible growers must apply recognised IPM techniques on a preventive basis. Non-chemical pest treatments are preferred over chemical treatments.'

(2)   Choice of chemicals: 'The crop protection product utilised must be appropriate for the control required. Growers must only use chemicals that are officially registered in the country of use and are registered for use on the crop that is to be protected, where such official registration scheme exists, or, in its absence, complies with the specific legislation of the country of destination. A current list of all products that are used and approved for use on crops being grown must be kept. This list must take account of any changes in pesticide legislation. Chemicals that are banned in the European Union must not be used on crops destined for sale in the European Union. Growers must be aware of restrictions on certain chemicals in individual countries.'

(3)   Records of application: 'All applications of pesticides must always include: crop name, location, date of application, trade name and name of operator. Pesticide application records must also include: reason for application, technical authorisation, quantity of pesticide used, application machinery used and pre-harvest interval.'

(4)   Pesticide residue analysis: 'Growers and/or suppliers must be able to provide evidence of residue testing. The laboratories used for residue testing must be accredited by a competent national authority to good laboratory practice (i.e. GLP or ISO 17025).'

Growers and grower organisations need EUREPGAP approval. This approval can only be achieved through independent verification by a national inspection or certification body that needs to be accredited (http://www.eurep.org, 6/9/2000). Full compliance of all 'musts' is necessary for approval. At first, the expectation was that the first growers would be approved around June 2000, but this was delayed because of unresolved legal issues concerning accreditation (http://www.eurep.org, 21 September 2000). All approved growers and grower organisations will have the right to use the EUREPGAP logo, as a means of communication in the business-to-business area and not (yet) designed to be used in the communication with the final consumer. The first approvals were expected before the end of 2001.

During the process of elaborating standards, EUREP has sought, and received, support from the European Commission. A clear sign of the Commission's interest is the fact that Community officials deliver speeches at EUREP conferences. In the future, it should not be considered unthinkable that financial support to farmers will be made dependent on compliance with EUREPGAP standards, and that the Commission will use the EUREPGAP protocol as a basis for Community framework legislation comparable to that for organic agriculture. Table 12.4 summarises the development of standard setting and certification for ICM.

Table 12.4 Development of standard setting and certification for ICM.

Year	Event
End 1970s	Introduction of integrated production
Start 1980s	Establishment of several national and regional initiatives
End 1980s	First ICM-labelled products
1997	Foundation of EUREP
2001	Publication of EUREPGAP protocol

Overview of the market situation for sustainably produced food

Organic production and consumption steeply increased in the last decade, especially since the second half of the 1990s. In our 1999 study focusing on the EU (van der Grijp & den Hond 1999), we saw that the number of organic farms rose from fewer than 10 000 to more than 80 000, and that the organic acreage increased from less than 250 000 to more than 2 200 000 hectares in ten years. The latest figures show that the growth of organic acreage is still continuing, and it is the general expectation that during the next years similar huge growth rates will be reached. Our study from 1999 showed that the main factors determining the level of organic production in a country are government support, especially subsidies for conversion; involvement of retailers and the food processing industry; consumer demand; and export potential.

For our study we combined the figures of relative organic acreage in 1998 and annual growth rates between 1993 and 1998 for fifteen European countries; we compared these with EU averages, and then categorised the countries into four main groups. Table 12.5 shows the outcomes of the categorisation.

The first group consists of countries with a high relative share of production as well as a high growth rate (both above the EU average). They may be considered the

Table 12.5 Categorisation of countries according to their level of organic production (van der Grijp &
den Hond (1999), and based on data in Rippin (1999)).

Relative share/Annual growth rate (1993–1998)	Above the EU average of 23.0%	Under the EU average of 23.0%
Above the EU average of 1.6%	*Booming countries:* Denmark Finland Italy	*Stabilising countries:* Austria Germany Sweden Switzerland
Under the EU average of 1.6%	*Countries with a high potential:* Greece Ireland Norway Portugal Spain	*Countries lagging behind:* Belgium France Luxembourg Netherlands UK

booming countries in organic production. Countries belonging to the second group
seem to have passed the 'booming' years and are now stabilising which is indicated by
a high relative share in combination with a low growth rate (below the EU average).
The third group consists of the countries with a high potential. They combine a low
relative share with a high growth rate. In a few years they may prove to be either
booming or lagging behind. The fourth group, not surprisingly, consists of countries
that are simply lagging behind which is represented by a low relative share as well as a
low growth rate.

As it is now in 2001, it seems that the 1999 qualifications about the relative position
of countries are still valid (see Table 12.5). A comparison shows that the same countries are still under and above the line indicating the EU average of the relative share
of organic production. In the past two years this average has risen from 1.6 to 2.94%
of the total agricultural area. There are, though, indications that some countries are
changing positions. Germany and the UK, for example, have experienced huge
increases of organic acreage during the past years. Table 12.6 provides an overview of
the five European countries with the highest organic acreage in absolute terms,
according to the most recent figures collected by SÖL in Germany (Schmidt & Willer
2001).

With regard to consumption, the market for organic food, though still small, is
growing at a fast rate, and is also forecast to grow further in the next decade. For our

Table 12.6 Top five of European countries with the
highest organic acreage on 31 December 2000 (based on
Schmidt & Willer 2001).

Country	Organic acreage	Relative share
1. Italy	1 040 377 ha	7.01%
2. Germany	546 023 ha	3.20%
3. UK	527 323 ha	2.85%
4. Spain	380 838 ha	1.49%
5. France	370 000 ha	1.30%

1999 study (van der Grijp & den Hond), we considered the value of 1% as the critical value above which the consumer market in a certain country has left its niche and has become mainstream. According to our study, five European countries had already passed the 'magical' line in 1998, including Austria (5%), Denmark (3%), Germany (2.5%), Switzerland (1.5%) and Sweden (1–1.5%) (data from Comber 1998). In addition, several countries were on the verge of a breakthrough having achieved a market share of 1%, including Finland, Italy, Norway and the UK.

In 1999, the International Trade Centre (ITC) published its study 'Organic food and beverages: world supply and major European markets', with the main aim to inform developing countries about the market potential of organic products. According to this study, annual growth rates of organic sales will range from 5–40% over the medium term, and in some major markets relative shares of 10% will be realised in the next few years. Table 12.7 provides an overview of the five European countries with the highest organic consumption in absolute terms in 1997, including a forecast for the year 2000.

Table 12.7 Top five of European countries with the highest organic consumption in 1997 (based on ITC (1999)).

Country	Organic consumption in 1997 (in $m US)	Forecast of organic consumption in 2000 (in $m US)
1. Germany	1 800	2 500
2. Italy	750	1 100
3. France	720	1 250
4. UK	450	900
5. Switzerland	350	900
Europe (total)	6 255	8 450

Up to now, the level of consumer interest in organic foods is generally higher in northern, western and middle European countries than in southern Europe, for a variety of reasons, including the emergence of food scares, the debate on GM crops, the better availability of organic products and the higher standard of living. The countries in the Mediterranean have developed especially strong positions as exporters of organic products, being the suppliers of the other European markets. Regarding imports and exports, The Netherlands has a rather exceptional position because of the activities of specialist companies that trade organic products from all over the world, then ship them to The Netherlands, and subsequently export them again.

The supply side of the European organic food market has always been highly fragmented, with thousands of small to medium-sized companies in operation (Comber 1998). More recently, however, there have been several important developments that are an indication of a radical restructuring of the organic market. In the first place, conventional supermarkets have become increasingly involved in the sales of organic products, and have launched organic retailer 'own brands'. It is noticeable that a high involvement of the supermarket channel usually coincides with domestic organic food consumption above or at the critical value of 1%. This is the case for Austria,

Denmark, Finland, Sweden, Switzerland and the UK. Interestingly, Germany, with its extensive network of nature food stores, represents the opposite situation as consumption is above 1% and supermarket involvement is under 50%.

In the second place, several conventional food processing companies have started to offer organic product lines besides their regular ones. Among these companies are some of the large multinationals, for example Groupe Danone (France), Del Monte (UK), Nestlé (Switzerland), and Unilever (UK/Netherlands). Practice shows that conventional companies generally choose between two strategies when entering the organic market: they either take over a specialist organic company or establish a new product line from scratch.

In the third place, interaction processes started up between the previously totally separate circuits of conventional and specialist organic companies. Several specialist producers and trading companies have started to supply conventional companies, especially the large supermarket chains. Some of these specialists changed their strategic behaviour by forming alliances and partnerships to ensure more consistent supplies, to improve access to distribution channels, and to benefit from economies of scale (see Comber 1998). Not surprisingly, specialist companies are now experiencing growth rates that far exceed those of earlier years.

Compared to the organic market, the documentation and statistics about the development of production and consumption under ICM are rather haphazard and, if available, only on a regional or country level. An extra complicating factor is the variety of definitions which are presently in use for ICM. The lack of figures may be remedied if a European-wide agreement can be reached about a baseline definition of ICM, and an internationally recognised inspection and certification system is introduced. However, it is without doubt that production and consumption figures under ICM are many times greater than those under organic protocols. Most consumers, though, are not aware that they buy ICM products because these products do not always bear an indication of their origin, and are usually sold for similar prices as conventional products.

Retailer initiatives in four European countries

When taking a closer look at the current initiatives of the food and retailing industry to market more sustainable food products, the picture varies between countries, sectors and companies, due to the nature of food production chains, economic driving forces, national governmental policies and cultural characteristics. To illustrate this variety, the following subsections go into more detail about specific retailer initiatives in four European countries. The selection of countries is arbitrary but they all have in common that in some way or another their national food industry is performing a pioneering role in the development of initiatives for more sustainable agricultural practices. Other selection criteria have been the different national contexts, and last but not least the availability of information.

Italy

Italy is one of the European countries with the highest increase of organic agriculture in recent years. Since 1997, it has in absolute terms the highest organic acreage in Europe (*cf.* Table 12.6). An important feature of the Italian market is its large export business, with sales being driven by a rise in demand for organic fresh produce in other European countries, especially Germany and France (Comber 1998). It is estimated that 60% of production is being exported.

The Italian domestic market for organic products is small but increasing, with most consumers living in the more affluent, northern part of the country. Nature food stores are the most important outlet for organic products in Italy, but super-markets have increased their market share in recent years (see Comber 1998; Zanoli 1998). They now manage 25% of organic sales (http://www.sana.it, 15 January 1999). Organic products are available in all major supermarkets, with market leader Coop Italia – a company with an overall pro-active environmental policy – as the most serious supplier. Euromercato is also significant in the organic market, while Conad, Esselunga and Gs stock some organic lines. However, the range of products available appears to be more limited than that in other countries.

The Italian organic sector has a large potential for further growth. In addition to the increasing export business, it is believed that a 4% domestic market share of organic products will be realised in the next few years.[3] However, the enormous increase of organic production in recent years is presently posing problems to the organisational and logistic structures available (Willer 1998).

Italy is also in the forefront with regard to production under ICM. Apofruit, one of the largest Italian growers' organisations, introduced its own trademark for ICM products 'Almaverde' as early as 1987. The retailer Coop followed with its own label 'Prodotti con Amore – Coop' in 1988, and is still the market leader for ICM in the country. In the years 1988–1997 the volume of ICM products sold by Coop increased from 6800 to 109 000 tons (EUREP 1998).

The Netherlands

The Dutch market for organic products is lagging behind in comparison to most other European countries. It has been suggested that the main reasons for this include the late introduction of organic products by the major supermarket chains, and consumer atti-tudes (see Comber 1998; Kortbech-Olesen 1998). Dutch people have the reputation of not spending too much money on food. EU statistics indeed show that Dutch consumers spend a smaller percentage of their income on food than most EU citizens.

Nature food stores have long been the most important retail outlet for organic products. In 1997, they still accounted for 75% of total sales (Comber 1998). Never-theless, in 1998, the balance began to turn with an increase of the market share of supermarkets: they now account for 35–40% of sales (Rabobank, press release 28 October 1998). From the three largest Dutch supermarket chains, Albert Heijn (owned by Ahold) is in the forefront in the development of an organic product offer.

Albert Heijn is the largest food retailer in The Netherlands with an estimated market share of 28% realised through its 670 shops. The company positions itself as a high-quality supermarket, attracting a relatively large share of the better educated, middle- and higher-income groups of the Dutch population. Up to 1997, Albert Heijn offered a selection of just twenty organic products. However, in February 1998, the company announced the launch of a new own-brand label ('AH Biologisch') which is to cover an extensive range of lines, including fresh produce, meat, dairy products and groceries. Reportedly, the company decided to extend the organic product range only after consumers expressed demand through a petition (HP/De Tijd, 17 July 1998). Subsequently, Albert Heijn asked its suppliers if they were able to supply organic products besides the conventional lines. Several suppliers reacted positively to the request and as a result the number of conventional companies involved in the organic market increased significantly.

To promote the organic product range, Albert Heijn uses its monthly customer magazine (*Allerhande*, 2.1 million copies). In a later stage, Albert Heijn intensified its promotion campaign with television advertisements, temporary price reductions and the publication of leaflets to inform consumers about the organic home brand. The company aimed to offer 500 organic products by 2001. Recently, Albert Heijn publicly expressed its concern about the high price level of organic products and announced a stronger collaboration with producers and processors with the aim of achieving more reasonable price levels (BFN News Service, 9 March 1999).

With regard to ICM, the retailer Albert Heijn started to implement an ICM programme ('Aarde & Waarde') back in the early 1990s. The publication of the Multi-Year Crop Protection Plan ('MJP-G') by the Dutch government was the impetus to start the programme. Albert Heijn had the advantage of its policy of long-term supply contracts. It was because of its concern for quality that the company has traditionally exercised a tight control over the various supply chains, which in the case of fruits and vegetables consist of a limited number of direct suppliers who buy the produce of selected farmers. The ICM programme started with the development of ICM standards to which Dutch suppliers and farmers should conform. Subsequently, the programme has been implemented step by step within the existing supply chain management system. At first a small number of farmers started with ICM on a limited number of crops. The number of crops along with the number of participating suppliers and farmers were increased such that in 2000 all crops from Dutch origin were produced under ICM standards. In 1996, Albert Heijn also started with an ICM programme for its foreign suppliers and farmers, beginning in Italy and Spain. It is Albert Heijn's final aim to offer all fresh produce, produced under either ICM or organic standards. In the context of the ICM programme, Albert Heijn cooperates with several other large European retailers in the EUREPGAP initiative.

The future of food industry initiatives for more sustainable agricultural production will be, more or less, affected by current policy developments in The Netherlands, as the government has published a strategic document called 'Integrated management, the way ahead, crop protection policy up to 2010' (*Zicht op gezonde teelt*). The central objective of the new policy is the realisation of integrated production on certified farms, ultimately in 2010, and it will be accompanied by a set of financial incentives. According to the government, the market, the production chain and individual growers are all responsible for the development and application of integrated crop protection. To give a framework to the objective of integrated management, the government will develop two different sets of ICM standards. The first set is an ambitious one, for farmers in the forefront, with innovative practices, and the second one will be less ambitious and is targeted at mainstream farmers. At this stage, the government has not taken any definite decisions about the level of standards to be applied.

Another relevant government document concerns a policy plan for promoting organic agriculture between 2001 and 2004 (*Beleidsnota biologische Landbouw*). Consistent with the previous policy document (*Plan van Aanpak Biologische Landbouw 1997–2000*), but unlike neighbouring European countries, the Dutch government holds the opinion that market forces will determine which share of organic production is feasible in The Netherlands. Considerable effort and money will be put into research, increased cooperation in organic product chains and tax incentives. A remarkable fact, though, is that the Dutch government, as the first in the EU, plans to abolish conversion subsidies after 2002. At a first impression, it seems that the Dutch government has chosen the option of ICM as the most realistic form of sustainable agriculture for The Netherlands, and that in its opinion every farmer converting to organic agriculture is a bonus.

Switzerland

Switzerland has experienced a 'bio-boom' since 1990 because of high government commitment and strong impulses from the market (Niggli 1998). The country has the second highest percentage of organically farmed land in Europe, after Austria. It has also a well-developed organic food market that is well ahead of that of most European countries. The two leading retailers, Coop and Migros, are playing a key role in the development of the organic market, and specialist nature food stores also constitute a major distribution channel (Comber 1998).

The company Coop started to sell organic products in 1993 (Niggli 1998). Migros entered the market in 1996 and, by now, all Swiss supermarkets are offering organic products. It can be seen that both Coop and Migros adopted a policy of 'regional preference' (Comber 1998). This means that, when choosing product lines, preference is given firstly to local products, secondly to products from neighbouring countries and regions, and lastly to imports.

Coop has by far the biggest sales of organic products, offering around 150 different products under its own label 'Naturaplan'. The company is responsible for almost half of all organic sales in Switzerland (Wehrle 2000). Thanks to Coop's successful range of organic food products, the company, being consistently second in the Swiss market, has been able to gain market share from its major competitor Migros. As a market leader, Coop plays an important role in the price-setting of organic products, with smaller competitors adapting to Coop's price level. At present, all Swiss retailers follow high pricing strategies that bear the full costs of the respective products (Wüstenhagen 1998).

As already mentioned above, Switzerland is the country of origin of integrated production. In 1978, a group of fruit producers using integrated production methods founded SAIO (*Schweizerischen Arbeitsgemeinschaft für integrierte Obst-produktion*). SAIO is an independent working group, and consists of researchers, technical consultants, retailers, wholesalers and producers. Among its activities are the drawing-up of guidelines and minimum standards for integrated production of fruit, which are published on an annual basis. The Swiss Fruit Association, SOV (*Schweizerischen Obstverband*), is responsible for the control of production under ICM. It introduced a national ICM label in 1990. In its first operational year, the labelling scheme only covered apples and pears, but its scope was soon extended to berries and vegetables, all of Swiss origin. Since 1998, the basis for certification has changed from the labelling of products to the certification of farm enterprises.

After the introduction of the SOV labelling scheme, the large retail chains introduced their own label lines for ICM products. The commercial success of these initiatives is mainly due to the following reasons (EUREP 1998):

- advantage of added value from an ecological point of view;
- increasing consumer concern about product quality;
- strict legal requirements regarding producer liability;
- perceived professionalism of producers under ICM protocols.

To date, the Swiss circumstances for organic and low-input agriculture may be considered rather ideal, and it is not inconceivable that future Swiss farming will all be either integrated or organic.

United Kingdom

Organic production in the UK is showing a steep increase after a slow start that was mainly due to a previous lack of government funding and support (Stolton 1998). The consumption of organic products was already on the rise. As early as the mid-1980s organic foods became available in supermarkets and these have been a dominant force ever since, representing a share of around 70% of sales (see Comber 1998; O'Hara 2000). Recent food scares and the debate about genetically modified organisms have stimulated the market (Comber 1998).

The first supermarket to introduce organic foods was Safeway in 1981, and by 1990 the 'big five' – Asda, Safeway, Sainsbury's, Tesco and Waitrose – were all stocking them (Comber 1998). The major supermarket chains all developed their own strategies to market organic products. Tesco, for example, offers organic produce at the same price as conventional lines, and is financially supporting a newly created organic agriculture research centre at Aberdeen University (Organic Food News UK, 29 September 1998). Furthermore, it works with The Soil Association, a certification organisation, to help develop the organic produce sector, and actively encourages farmers to move into organic growing. Tesco now offers more than 700 organic products (Norton 2000).

It is Waitrose's strategy to replace conventional produce with organic produce wherever possible. Waitrose offers the largest range with more than 1000 products (Norton 2000). Asda now offers more than 400 organic products and is going to invest heavily in a new range of organic products (Norton 2000). Sainsbury's is presently the retailer with the highest organic market share, notably 2.5% of its general sales (Haest 2000). The company has put a strong emphasis on the development of its organic own label. Sainsbury's introduced a range of initiatives to encourage the development of the organic market; for example it established SOURCE, the Sainsbury's organic resourcing club which works together with the Soil Association to find new organic suppliers, and recently Sainsbury's announced a labelling scheme for products in the process of conversion, which will display the phrase 'produced under conversion to organic farming'.

Iceland, which specialises in frozen foods, occupies a special place in the British retailer landscape, although it only has a market share of less than 2% (Marsden *et al.* 2000). Iceland has become well-known for its pioneering stance on food issues, by which it tries and defines its place in the market (see Marsden *et al.* 2000). One of its previous initiatives was to guarantee that no Iceland own-brand product manufactured after the date of 1 May 1998 would contain any genetically modified ingredients. A later initiative implied a complete switch to products from organic origin, starting with its own-label frozen vegetables, and other conventional foods following soon after. Iceland took the decision to charge no premium for organic products, and to offer them at a similar price as that for non-organic ones. To reach that goal, the company planned to invest £8 million by way of reduced profit margins (http://www.iceland.co.uk, 20 June 2000). Initially Iceland claimed it had secured contracts to buy in almost 40% of the produce grown organically world-wide (Agrarisch Dagblad, 16 June 2000). At a later stage, however, the company had to admit that its plans had been too ambitious and that expectations had to be adjusted to a lower level.

The large English supermarket chains also made progress with ICM programmes. The National Farmers Union (NFU) formed a partnership with a group of large retailers (ASDA, CWS, Sainsbury's, Marks and Spencer, Safeway, Somerfield and Waitrose) which aimed to establish a scheme based upon ICM principles for fresh fruits and vegetables of English origin. This so-called Assured Produce initiative, which started in 1991, seeks to achieve its objectives in three phases:

(1)   to establish for each crop a baseline of current best horticultural practice;
(2)   to verify independently that growers are reaching their standards; and
(3)   to lift these standards measurably.

Initially the crop-specific protocols described existing best agricultural practice, but it was the intention to update them annually following a formal review of new developments. The first protocols were published in July 1993. In addition to the Assured Produce protocol, participating retailers may wish to adopt extra or higher standards. Tesco, for example, had already initiated its own programme ('Nature's Choice') prior to the formation of the partnership between the NFU and the large retailers, and has continued its development.

The British retailers participating in the Assured Produce scheme have subsequently taken the lead in a Europe-wide initiative with the aim of imposing similar standards on overseas suppliers. It is a noticeable fact that the resulting EUREPGAP protocol has given British farmers' organisations cause to deliver a complaint to the European Commission against the Assured Produce scheme. They claimed to face requirements that are stricter than those for non-UK farmers. The organisations would favour the Assured Produce scheme being adjusted to the level of the less demanding EUREP scheme. It seems now that the objections about the Assured Produce scheme have been taken away, as the chairman of Assured Produce officially declared that 'Assured Produce is benchmarking itself against the EUREPGAP framework so that UK growers and consumers can benefit from a level playing field in relation to competition from abroad' (press communication, Assured Produce, 9 July 2001).

Bottlenecks and challenges

As was already mentioned in the first section, our institute has conducted a research project that focused on initiatives with an impact on pesticide use that are taken by the Dutch food industry (van der Grijp *et al.* 2001). During the project we identified bottlenecks that companies experience in moving towards sustainability and explored opportunities for improving the interactions between private initiatives and public policy. We collected research material through a survey among Dutch food companies, a series of in-depth interviews, and a round-table discussion with participants from the food industry, the government and social organisations. In this section some of the preliminary results from this research will be presented, followed by more general remarks about bottlenecks and challenges.

According to the IVM research, most Dutch companies are perceiving bottlenecks in the development of initiatives for a product offer based on sustainable agricultural production. This counts more strongly for the companies with an organic initiative than for those with an initiative for integrated production. Table 12.8 shows that in particular the tuning of supply and demand creates a number of problems. It can be concluded from the table that many companies perceive the supply of products from sustainable agricultural production as too small or too irregular, and demand as smaller than expected. In the interviews these perceptions about the mismatch of supply and demand were confirmed but also put into a more balanced perspective. There seems to be a connection with the place a company occupies in a specific food supply chain; retailers will perceive supply as relatively more problematic than will, for example, traders. With regard to organic initiatives, mismatch of supply and demand has also been reported by other researchers (see Michelsen *et al.* 1999).

The price level of organic products is another bottleneck that is considered as seriously hampering the development of the organic market. Premium prices to be paid for organic products may be very high compared to prices for conventional products. Interestingly, Michelsen *et al.* (1999) found that there is not always a connection between premium prices asked for in shops and price premiums paid to farmers. In the UK, there have been public disputes about price differentials between organic and conventional food products. British supermarkets have been accused of wielding too much power and exploiting demand by elevating prices (O'Hara 2000). In return, they

Table 12.8 Bottlenecks in the development of initiatives for organic and integrated production in The Netherlands.

Bottlenecks	Organic initiative (*n*=22)	Initiative for integrated production (*n*=10)
Supply is too small	55%	40%
Prices are too high	50%	0%
Development of guidelines takes time	—	30%
Conversion of farmers takes time	36%	20%
Supply is too irregular	32%	30%
Product properties are deviant	32%	—
Demand is too small	23%	20%
Willingness to convert is too small	14%	10%
Separate production facilities are too expensive	9%	10%
Regular production capacity is too big	9%	—

have claimed that their margins on organic foods are identical to those on conventional produce, but that the higher costs are due to the organic production system and the high level of imports (75 %).

The ambition level of organic and ICM standards varies between countries. For example, some countries are allowed to use certain substances that are banned in other countries. Even more pressing is the risk of fraud, because it is not always possible to assess what is actually happening in the field, or in the other stages of food processing, packaging and distribution. The great danger is that if fraud is discovered to be on a relatively large scale, consumer confidence will be severely damaged. For organic products this may result in a decreased willingness to pay premium prices and eventually a collapse of the organic market. More susceptible to fraud are especially those countries which have less stringent inspection and certification systems in place. In this context it is one of the big challenges for supermarkets to maintain credibility by guaranteeing a high integrity of production systems. Therefore, supermarkets are presently improving the traceability of products, for example by developing identification systems and establishing long-term relationships with farmers and other suppliers.

With regard to organic products, there is also a fear that their quality will suffer as they become mass-produced, and that the demands of big business will gradually undermine the original aims of the organic movement. There are already examples of standards being adapted to the wishes of the market. In contrast to this erosion of standards, however, there is also a tendency to heighten the ambition level of organic standards by the development of organic-plus systems. The retailer Sainsbury's and the food processing company Heinz, for example, have chosen this more ambitious approach.

As was already described above, standard setting for ICM is in an earlier stage of development than that for organic agriculture. In the next few years the challenge for

EUREP and its members will be to promote and support the implementation of the EUREPGAP requirements in national and regional standards for ICM. This means that countries have to develop activities which will enable their growers and growers organisations to become business partners of the EUREP members. These partnerships may be essential in future supply chain arrangements. For the countries that already have standards for ICM in place, it will be necessary to evaluate whether their systems comply with the bottom line set by EUREP. Countries that do not have any standards for ICM should make an effort to create compatible systems.

Future perspectives for market-driven initiatives

It is evident that an ever-increasing number of retailers and food processing companies are developing supply chain initiatives that define and implement more sustainable agricultural practices. In return, these initiatives seem to be for farmers a convincing impetus to change their agricultural production methods, and are thus instrumental in reducing farmers' dependency on pesticide products. It goes beyond the scope of this chapter to draw conclusions about the extent of the reduction in pesticide use that directly results from food company initiatives. However, it seems not to be a far-fetched assumption that these initiatives have indeed a not negligible effect on the behaviour of farmers with regard to pesticides. This does not necessarily imply that 'conventional' farmers become less dependent on pesticides. At best, one could hope for a 'trickle-down' effect of alternative production techniques. In addition to the direct effects on farmers' behaviour, food industry initiatives are also proving to be a source of inspiration for governments, at EU and national as well as at local level, to come up with supporting measures. Examples include the interest of the European Commission in the EUREP initiative, and the new Dutch policy document on pesticides that also targets food industry stakeholders.

The factual information in this chapter can be largely characterised as a 'snapshot' at a given moment in time. As developments in the food industry are rapid and even accelerating, the picture presented here may radically change in the next few years. In general, though, it may be concluded that there are the following perspectives for the future:

- The large retailers will obtain an ever-increasing market share and market power.
- The food industry will put ever more emphasis on quality and safety aspects of products.
- The food industry will aim at increased standard setting, certification and product traceability.
- EU policy will increasingly provide incentives to convert to more sustainable agricultural practices, directly by the reform of agricultural policies and indirectly by more stringent liability schemes.
- ICM as defined by EUREPGAP will become the general standard for the production of fruits and vegetables in Europe.
- Organic products will become mainstream.

It seems that present developments are pointing in the right direction really to achieve reduction of pesticide use. To speed up the process, it is recommended that governments should give extra impulses by developing supportive policies. The European Commission, for example, should devise legislation defining the bottom line for ICM, and should make financial support to farmers dependent on compliance with ICM standards. National governments should facilitate the development of schemes aimed at standard setting and certification of organic as well as integrated production. The ideal would be to create schemes that also provide opportunities and incentives for growers in the front line. In addition, governments should assist in creating a good infrastructure for education, research and knowledge dissemination. Last but not least, there is a great need for consumer-awareness campaigns based on well-balanced information about the different types of sustainable agriculture and their relation to food quality and safety, the environment and health.

Notes

1 Council Regulation 2092/91/EC, OJ L198, 22 July 1991.
2 Council Regulation 331/2000/EC, OJ L48, 19 February 2000.
3 Statement of Sergio Rossi, general manager of Sana, made at Bio Fach 1999, 19 February 1999.

References

Browne, A.W., Harris, P.J.C., Hofny-Collins, A.H., Pasiecznik, N. & Wallace, R.R. (2000) Organic production and ethical trade: definition, practice and links. *Food Policy*, 25, 69–89.

Comber, L.R. (1998) *The European Organic Foods Market*. Leatherhead Food RA.

EUREP (1998) *Integrated Production of Fruit and Vegetables. Including 9 Country -specific Case Studies*. EUREP, Cologne.

Green, K. (2000) Sustainability of food consumption and production systems. *IHDP Update*, (2), 1–3.

van der Grijp, N.M., & den Hond, F. (1999) *Green Supply Chain Initiatives in the European Food and Retailing Industry*. Research Memorandum 99/07. IVM, Amsterdam. (http://www.vu.nl/ivm)

van der Grijp, N.M., de Boer, J. & den Hond, F. (2001) *Initiatieven vanuit de Nederlandse voedingssector ter beperking van het bestrijdingsmiddelengebruik*. Research Memorandum 01/03. IVM, Amsterdam. (http://www.vu.nl/ivm)

Haest, C. (2000) EKO in de Europese supermarkt. Presentation at the congress *De groeistuipen van EKO*, Noordwijkerhout, 7 March 2000.

Hertzberger, S.A. (1999) Duurzaamheid in de voedingsmiddelenketen. Notitie ter gelegenheid van de eerste DuVo-dialoog op 9 december 1999, Rotterdam.

ITC [International Trade Centre] (1999) *Organic Food and Beverages: World Supply and Major European Markets*. ITC/UNCTAD/WTO, Geneva.

Kortbech-Olesen, R. (1998) Export potential of organic products from developing countries.

Paper presented at *IFOAM conference, 1st International Seminar on Organics in the Super-market*, Mar del Plata, 14–15 November 1998.

Marsden, T., Flynn, A. & Harrison, M. (2000) *Consuming Interests. The Social Provision of Foods*. UCL Press, London.

Michelsen, J., Hamm, U., Wynen, E. & Roth, E. (1999) *The European market for organic products: growth and development*. (Organic Farming in Europe: Economics and Policy, Volume 7.) Hohenheim.

Morris, C. & Winter, M. (1999) Integrated farming systems: The third way for European agriculture. *Land Use Policy*, 16 193–205.

Niggli, U. (1998) Ökologischer Landbau in der Schweiz. In: *Ökologischer Landbau in Europa. Perspektiven and Berichte aus den Ländern der Europäischen Union und den EFTA-Staaten* (ed. H. Willer), pp. 332–49. Deukalion Verlag, Holm.

Norton, Ch. (2000) Organic food is a 'waste of money'. *The Independent*, 2 September 2000.

O'Hara, M. (2000) News analysis: Organic farmers fear growing pains: demand for traditional food values comes into conflict with 21st century mass marketing. *The Guardian*, 25 July 2000.

Perkins, J.H. (1982) *Insects, Experts and the Insecticide Crisis*. Plenum Press, New York.

Rippin, M. (1999) *Materialien zur Marktberichterstattung. Strukturdaten zum Ökologischen Landbau*. Sonderdruck zur BIOFACH '99. ZMP, Cologne.

Schmidt, G. & Willer, H. (2001) Organic farming in Europe: Provisional statistics 2001 (http://www.organic-europe.net, 8/10/2001).

Stolton (1998) Ökologischer Landbau in Grossbritannien. In: *Ökologischer Landbau in Europa. Perspektiven and Berichte aus den Ländern der Europäischen Union und den EFTA-Staaten* (ed. H. Willer), pp. 148–69. Deukalion Verlag, Holm.

Wehrle, Th. (2000) Organic products in Swiss supermarkets. Paper presented at *IFOAM Conference, 2nd International Seminar on Organics in the Supermarket*, Basel 25–26 August 2000.

Willer, H. (1998) Einmaliges Wachstum des Biosektors in Italien. (http://www.soel.de/infos/internat/italia/italia98.htm)

Wüstenhagen, R. (1998) Pricing strategies on the way to ecological mass markets. Paper presented to the *Greening of Industry Network Conference*, Rome, 15–18 November 1998.

Zanoli, R. (1998) Ökologischer Landbau in Italien. In: *Ökologischer Landbau in Europa,. Perspektiven and Berichte aus den Ländern der Europäischen Union und den EFTA-Staaten* (ed. H. Willer), pp. 198–217. Deukalion Verlag, Holm.

Chapter 13
Impact of International Policies (CAP) and Agreements (WTO) on the Development of Organic Farming

David Barling

Introduction

Macro-level agricultural policies, such as the Common Agricultural Policy (CAP) and international trade agreements, as agreed under the General Agreement on Trade and Tariffs (GATT) Uruguay Round, provide regulatory frameworks within which farmers in the European Union (EU) take decisions about production. The CAP has been a major determinant of agricultural production in the EU. Since 1994, with the completion of the GATT Uruguay Round agreements, it has had to operate within an international rules-based trading regime for food and agriculture under the auspices of the World Trade Organisation (WTO). The most important of these agreements was the Agreement on Agriculture (AoA), which sought to liberalise food trade and reduce state supports for farming. The EU had responded to the international trade negotiations around the GATT in the late 1980s and 1990s, with two phases of CAP reform: firstly, the so-called McSharry reforms of 1992, in preparation for the completion of the Uruguay Round, and subsequently the Agenda 2000 reforms. The latter were also in response to the EU's planned enlargement of its membership to Central and Eastern European states, as well as the AoA. The AoA had a built-in review process that started in 2000, and will further impact upon the CAP and its ongoing reform. The CAP, in turn, has a mid-term review of some of its own reforms scheduled for 2003.

The Uruguay Round was an endorsement of the liberalisation of international trade, as advocated by large industrial and corporate interests and leading industrial nations. This so-called 'Washington consensus' came to dominate world politics (Thomas 1999; Lang & Hines 1993). The advance of this liberalisation (or neo-liberal) agenda reflected the economic thinking that held that competition on the world food market would stimulate agricultural production systems to maximise production at minimum direct costs, and enhance further economic growth (House of Commons 2000a). Conversely, the liberalisation agenda and the Uruguay Round agreements have been criticised for paying insufficient attention to the external environmental, social, and animal welfare costs of agriculture and its sustainability (Crompton & Hardstaff 2001; Mander & Goldsmith 1997).

The dominant model for agriculture practised in the United Kingdom, and much of western Europe after the end of World War II became one driven by the economic priorities of the state for self-sufficiency in food production. The further industrialis-ation of agriculture, increasingly capital- and chemical-intensive, led to increases in farm size, external technological inputs, and yield and food output. The effects of this process led to increasing concern over environmental, animal welfare, food safety, health and social impacts. For the UK, entry into the European Economic Commu-nity and the CAP consolidated this pattern of state support for agricultural production. A less desirable consequence was overproduction of certain foodstuffs such as dairy products, beef and cereals, adding budgetary pressures to a potential reform agenda of the CAP. The drive to increase production also lead to rationalis-ation within farming, with technical inputs replacing labour and concentration of farm holdings leading to the subsequent reduction of numbers of farmers and farm workers and so the dislocation of rural communities and of social sustainability (Pretty 1998; Whitby 1996). Social theorists of agriculture have conceptualised such developments as part of an industrial model of agriculture (Goodman *et al.* 1987) or a productionist model (Lang 1999). Certainly, policymakers have come to characterise such a model as the pre-eminent or 'conventional model of agriculture' (House of Commons 2001). This chapter focuses upon the case of UK agriculture and organic farming within the context of the CAP, and subsequently the AoA.

In response to concerns about the adverse impact of the conventional model, a variety of approaches to farming have emerged. On the one hand, conventional agri-culture has adopted management strategies to mitigate specific problems. Strategies to reduce pesticide use at the farm level, for example, have led to the introduction of integrated crop management and pest control techniques. On the other hand, alterna-tive agricultural systems have been developed, under a variety of terms, advocating more ecological and post-productionist strategies (Beus & Dunlap 1990). Foremost amongst these alternative approaches, in Europe, is organic farming. The characteris-tics of organic farming are discussed in more detail in the next section, but amongst its more sustainable practices is the absence of the use of chemical or synthetic pesticides. Some natural pesticides are permitted, including highly toxic copper sulphate, although this is due to be phased out from use. Organic farming does provide agricul-ture with a clear alternative strategy to current pesticide use.

The key questions that emerge from this chapter are the extent to which the CAP (including the recent Agenda 2000 reforms) and the introduction of the AoA have impacted upon the development of organic farming, as an example of an alternative system of agriculture offering a more sustainable approach; also, to what extent might the review of the AoA and the planned review of the CAP offer further opportunities to organic farming? What reforms might be proposed for these regimes to promote organic farming as a sustainable alternative? Within this developing framework of multilevel governance, key policy decisions that shape the future of organic farming are being decided. Multilevel governance pictures a system of negotiations occurring at different levels while each level interacts with and conditions the discussions of the other. The framework is both horizontal (across different governments) and vertical. The supranational and intergovernmental regimes of the CAP and WTO provide systems of rules and norms within which these negotiations are framed and processed (Lang *et al.* 2001).

Organic farming has gained acceptance in dominant policymaking circles as an agricultural system that can both extensification of farming and environmental benefits. Organic farming organisations have had to downplay their role as a radical alternative to conventional agriculture in order to gain this acceptance. However, as the disadvantages of conventional agricultural practice have become more evident, so the impetus has grown for a more wide-ranging policy towards agriculture, going beyond a narrow productionist approach. The adoption of such a policy has been incremental, the latest phase being the acknowledgement, at least in the EU, that the role of agriculture is multifunctional, and can be the provider of wider public goods. It is within this formulation that policy change more conducive to organic farming may emerge at the EU within the CAP, and ultimately at the global level under the terms of the AoA. Such change is still in its very early stages and remains far from certain. It is also within such policy change that a singular outcome of organic farming, an end to synthetic pesticide use (and one focus of this volume) may increasingly occur in world farming. The characteristics of organic farming, as a system for producing more sustainable agricultural objectives, need to be more fully delineated first within the context of the contemporary agro-food system.

Organic farming, sustainable agriculture and the modern agro-food system

A related theme within both conventional and alternative agriculture over the past decade or more has been the need to make agriculture more sustainable. Sustainability has proven to be a universally embraced and appropriated concept, and consequently its meaning is highly contested. Jules Pretty (1998), as an advocate, has identified some key principles for sustainable agriculture; firstly, a thorough integration of natural processes such as nutrient cycling, nitrogen fixation, soil regeneration and pest–predator relationships into agricultural production processes; secondly, a minimisation of external and non-renewable inputs that damage the environment or farmers' health; thirdly, full participation of farmers in problem solving and greater use of farmers' knowledge and experience in seeking technical and technological solutions. Lastly, wildlife, water, landscape and other public goods of the countryside should be enhanced in terms of quantity and quality. Sustainable agriculture needs to draw upon, and in turn sustain, both natural capital and social capital. Natural capital (including soil, water, air, plants, animals and ecosystems) has to be integrated in agricultural systems in the form of regenerative technologies such as: use of nitrogen-fixing plants for soil conservation, use of natural predators for pest control and integration of animals into cropped systems. Social capital entails utilising farmer and community labour, knowledge and experience and underpinning community cohesion.

The importance of greater reliance upon natural organic inputs, farmer participation and reduction of external and non-renewable inputs is also reflected in the development of agroecology as a model for poor farmers in developing world agriculture (Altieri 1995; Pinstrup-Andersen *et al.* 1999). Similarly, the United Nations Food and Agriculture Organisation (FAO) in concert with the Convention on Biological Diversity has become an advocate for the importance of an agro-ecosystems approach in developing countries, embracing many of these principles of sustainable agriculture (Aarnik *et al.* 1999).

Notwithstanding the pre-eminence of the conventional agriculture model as presented, farming within the EU includes a wide diversity of practices, with a large amount of land subjected to more environmentally benign low intensity and traditional farming practices. A nine-country study from within the EU estimated that 38% of usable agricultural land (about 56 million hectares) is farmed with low-intensity and traditional practices, mostly in so-called marginal agricultural land and more remote, sparsely populated areas, yet providing important countryside and wildlife habitat conservation (IEEP & WWF 1994). Hence, the conventional model is more muted in its application in such areas.

The general impulse towards more sustainable practice within European agriculture might suggest a continuum of approach from conventional highly intensified systems to more environmentally aware integrated farm management systems to low intensity farming through to organic agriculture. Yet, it can be misleading to adopt such a linear approach. In the case of organic farming, a more holistic approach is advocated, setting it apart from such linear linkages. Organic farming sees 'the farm as an organism, in which all the component parts – the soil minerals, organic matter, micro-organisms, insects, plants, animals and humans – interact to create a coherent, self-regulating and stable whole', rather than in terms of external inputs (Lampkin *et al.* 1999).

Organic farming has its origins in the first half of the twentieth century, gaining momentum from the 1940s. There are acknowledged to be different pioneers across different European nations who developed their philosophy and principles under different titles. These included: Steiner and Pfeiffer and biodynamic farming originating in Germany and spreading across Northern Europe (including the UK); Hans Peter Rusch and Hans Muller and biological farming in Switzerland; and Albert Howard and Lady Eve Balfour and organic farming in the UK. In the UK there was early evidence of an evolution in thinking concerning organic farming. Balfour further developed Howard's work on soil management and his advocacy of adapting agricultural research to local farmer knowledge and local ecosystems into 'a powerful holistic approach linking the soil and its fertility to questions of animal and human health' (Clunies-Ross & Cox 1994). For a new generation of converts to organic farming in the 1960s and 1970s the emphasis upon locality, of local production for local markets, provided an attractive socio-economic alternative to conventional agriculture and the dominant agro-food system.

The standards of organic farming, derived from such principles, have not been immutable or set in scientific stone, but have evolved and adapted. Synthetic pesticides and genetically modified organisms (GMOs) were introduced into the farm system subsequent to the early development of organic farming, but those setting standards for organic farming have formulated responses to them, prohibiting their use in each case. Today, a system based on soil management and mixed farming of livestock and crops needs to formulate standards for quite different areas such as fish farming and complex food processing.

Organic producers have sought to establish their own standards and certification, a quasi- or private regulatory system, that provided market credibility and branding while putting up potential entry barriers (Guthman 1998). The state has welcomed such voluntary standardisation by the private sector. However, with the UK state highly engaged in food production, it sought to rationalise the regulation of organic standards, and set up the UK Register of Organic Food Standards (UKROFS) in 1987 to co-ordinate the setting of standards amongst the differing producer organisations. Regulation under the EC followed in 1991. According to Regulation EEC 2092/91 organic farming is defined as:

> a system of managing agricultural holdings that implies major restrictions on fertilizers and pesticides. The method of production is based on varied crop farming practices, is concerned with protecting the environment and seeks to promote sustainable agricultural development. It pursues a number of aims such as the production of quality agricultural products which contain no chemical residues, the development of environment-friendly production methods avoiding the use of artificial chemical pesticides and fertilizers, and the application of production techniques that restore and maintain soil fertility.
>
> (CEC 1991)

These EU-wide standards were amended in 1999 to cover livestock. Under the EU regulations the role of UKROFS has become one of 'ensuring that organic certifying bodies correctly interpret and implement [the Community] legislation rather than actually setting standards' (House of Commons 2001). Nonetheless, the EU regulations are the result not of any concrete science but are 'an embodiment of traditional practice more than anything else' (House of Commons 2001).

Organic farming has evolved in its organisation in response to market demands and opportunities and emerging research and development. The organisation of organic farming, particularly at the stages which are off, but near, the farm, is evolving to meet market demands in terms of distribution, wholesaling and marketing.

This has led to concerns that organic farming and food production is taking on the characteristics of industrial or productionist agriculture, compromising its integrity as an alternative agricultural strategy for food production. This is a critique levelled by an analysis of the organic vegetable commodity chain of northern California where the organic food market, which is more mature that that of the UK, provides examples of agri-business strategies. Examples included the large-scale monocrop production of carrots (regulations in California relate to inputs, such as pesticides, not processes, such as crop rotations) across a variety of different climatic locations on a contract supply basis (Buck *et al.* 1997). This has led social theorists such as Watts & Goodman (1997) to reflect on corporate dominance and to question the depiction of organic agriculture in terms of small producers and alternative networks. Increased demand for organic products has gone beyond fresh farm produce to include an increasing diversity of goods processed and manufactured from organic products. The pressures of industrial processing and manufacturing, as well as the intervention of multiple food retailers, have come to bear on organic farming and organic foods. The EU's regulation of organic processing standards has been less strict than that of organic farming. Concern has been raised over the number of non-organic processing aids and additives allowed, and the rules governing complex processing plants which make both organic and conventional foods using a contin-uous process (House of Commons 2001). The ingredients of the final manufactured food product may compromise the healthy food goal, and image, promoted by organic farming (Lobstein 1999).

Organic farming is trying to practise within agro-food systems affected by the contemporary pressures of globalisation, irrespective of the local and regional food economy-centred aspirations of some practitioners within the organic movement. In 1999–2000 an estimated 75% of organic food sold in the UK market was imported (Soil Association 2001a). The integration of organic food into large-scale and highly concentrated retailing was reflected in the UK organic sales through different outlets in 1999–2000. Supermarkets accounted for 74% of sales (up from 69% in 1998–1999), the independent retailers and health food shops sold 13% (down from 15%), and another 13% of products were sold through direct marketing schemes to consumer sources such as farm-gate sales, box schemes or market stalls (down from 15%) (Soil Association 2001). The more radical view of the role of organic farming as an alterna-tive food system, linking producer directly to consumer, is stymied in part by the lack of local and national production and infrastructure to meet market demand.

The UK's retail market for organic food had rapidly expanded in the UK through the 1990s. Sales had been worth £105 million in 1993–1994; £140 million in 1996–1997; £390 million in 1998–1999 and £605 million in 1999–2000 (Soil Association 2001a). Germany remains the largest and most mature organic market in the EU (£1.6 billion in 2000) representing 25% of the total annual organic turnover. Over the next few years from 2000, further growth of 20% is estimated in Italy (from an estimated £600 million in 2000), France (from £500 million), Sweden (not available), The Netherlands (from £145 million), and 40% growth in Denmark (from £240 million) (Soil Association 2001a). Such growth in market demand can be explained by consumers' increasing lack of confidence in the safety and overall impact of conventional food production (in both health and environmental terms) and the search for an alternative. Various food scan-dals and uncertainties, from salmonella to BSE, from GMOs to dioxins, have fed negative perceptions of conventional agriculture. Organic food has been able to promote an alternative 'brand image' for environmentally sound food production, and as a more 'natural' and 'healthy' food, free from pesticide residues. Multiple food retailers have bought into and further supported the success of this branding, seeking contracts with producers, offering shelf space and even funding conversion, in order to help meet market demand for these type of products (van der Grijp in this volume). In turn, such promotion has led to a renewed scrutiny of this image by the conventional agricultural and food establishment. The head of the UK's Food Standards Agency criticised the safety and health claims of organic food, prompting a detailed rebuttal from organic producers (FSA 2000; Soil Association 2001b).

The post-War state, in the UK at least, has supported the development of the dominant industrial model for agriculture. Despite the variety of participants within the organic movement, organic producers who have sought to gain the support of the state in order further to develop their own systems of food production have had to seek entry into this dominant and somewhat contradictory paradigm. The result for many organic farming organisations has been a strategy of seeking to avoid directly challenging the dominant model of agriculture. This strategy has entailed down-playing the alternative or holistic challenge to the dominant model, adopting a more pragmatic and incremental approach to gain acceptance for organic farming in the wider agricultural policy community. In the 1980s key alliances were made with the supermarkets and the agricultural training and education institutions, further pressuring the government departments to support relevant research (Clunies-Ross & Cox 1994). The initial price of such entry was the acceptance of organic farming in the 1980s by the Ministry of Agriculture, Food and Fisheries (MAFF) and the National Farmers Union (NFU) as one possible solution to overproduction under the CAP. The result was 'the persistent presentation of organic farming as an extensification option' (Clunies-Ross & Cox 1994). Similar thinking was reflected, for example, in a European Commission document in the early 1990s which presented organic farming as most suitable for marginal agricultural areas (already low impact and extensive) where premium pricing would allow farmers to be profitable (CEC 1994).

The terms under which the utility of organic farming as an add-on to conventional farming evolved somewhat further in the 1990s through the introduction of the European agri-environment regulation (EEC/2078/92). Increasingly, the environmental benefits of organic farming are being acknowledged, gaining some official acceptance in the UK in MAFF, and more positive advocacy in the Department of Environment (DETR) (Meikle 2000). These government departments were merged, in part, as the Department of Environment, Food and Rural Affairs (DEFRA) after the General Election of 2001. Advocacy for organic farming emerged in parts of the European Commission, also (CEC 1996). In June 2001 the Council of Agriculture Ministers invited the Commission to analyse the possibility of a European Union action plan to promote organic food and farming and to present appropriate proposals (Council of Ministers 2001).

The promotion of organic farming as a provider of public goods, in terms of enhancement of biodiversity and conservation of the soil and of the traditional landscape, has found increasing currency in policy debates surrounding the possible reform of farming and food production (Azeez 2000). Conversely, this has not prevented state policymakers, entrenched in the conventional agriculture paradigm, from advocating selective adoption of organic practices to improve the environmental management of that paradigm (see House of Commons 2001). Once again, organic farming as a holistic alternative is largely ignored. The promotion of these public goods does provide, however, a key entry point for organic farming into the reform debates surrounding the CAP and the AoA. The structure of organic farming under the CAP and its recent reform, and the potential for further reform, are addressed in the next section.

The CAP, Agenda 2000 reform and organic farming.

The state of European agriculture and rural life is largely shaped by the CAP regime, which has provided a macro policy framework, in a somewhat contradictory and imperfect way. The CAP also provides the umbrella under which member states can support organic farming. The aims of the CAP as set out in the Treaty of Rome (article 39) set the tone for the contradictions within this policy. The aims included to: increase agricultural productivity by promoting technical progress; ensure a fair standard of living for the agricultural community by increasing the individual earnings of persons engaged in agriculture; stabilise markets; assure availability of supplies; ensure supplies reach consumers at reasonable prices (CEC 1958). In short, the CAP was concerned not only with food production, but also with securing a degree of social stability in Europe during the post-War transition from a largely rural, peasant population to a largely urban one. The result has been a policy that has increasingly supported and subsidised European farmers and first-stage processors, and has been shaped by the realities of intergovernmental bargaining on behalf of national producer interests, at considerable costs (Grant 1997).

The basic instruments of the CAP in the past have been the purchase of produce from farmers to maintain prices (intervention buying), import levies to keep out other countries' produce, and export subsidies to compensate for selling on the world market at lower prices (Grant 1997). The negative impacts of the CAP have included a steady reduction in the number of farmers, particularly smaller farmers, an increase in farm size and concentration, unequal distribution of funds to larger farms (Podbury 2000), continuing rural depopulation and higher prices for food consumers. In addition, costs in terms of environmental damage, notably through production-enhancing artificial fertilisers and pesticides, as well as animal disease and its effects along the food chain, have been immense if not officially accounted. The costs of these 'hidden' externalities to UK agriculture alone were estimated at some £2.343 billion in 1996 (Pretty *et al.* 2000).

Efforts to ameliorate this impact have been made in an incremental manner. The McSharry reforms were the first significant attempt at reform since the introduction of dairy quotas in 1984 and budgetary reforms of 1988. The aims of the McSharry reforms of 1992 were to reduce production surpluses and to prepare European agriculture for entry into the GATT. The reforms represented a shift from price support for production to production controls and direct payments to farmers as compensation for the price support reductions. The reforms were aimed mainly at cereal and arable producers. Payments included: hectare compensation payments for set-aside (paying farmers not to produce on up to 15% of their land); headage payments for livestock producers and payments for livestock to compensate for poor climate (less favoured areas schemes); and hectare payments for altered systems of farming. The last included the introduction of the agri-environment measures (EEC/2078/92), the first significant environmental feature within CAP, although they only came to make up 2% of the total CAP budget.

The most recent reforms were Agenda 2000, agreed at Berlin European Summit in March 1999, which sought to address both the demands of the AoA and preparations for enlargement of the EU to include countries of Central and Eastern Europe (CEEC). Overall, the reforms are small and incremental. Opinions on the longer-term significance of the reforms are somewhat divided between those who see it as signalling very little change (*e.g.* Swinbank 2000a) and those who see the reforms as opening the door to future more significant reform (*e.g.* Tangermann 2001). Overall, direct payments to beef and arable farmers are increased while price support is reduced. In the case of cereals there is a reduction in the intervention price of 15% by 2002 but no further reduction until 2006. These cuts are compensated for by compensation equivalent to at least 50% in the form of direct payments for loss of income. In the beef sector there is a cut of 20% in the intervention price by 2002, alleviated by increases in the various premiums for example the slaughter premium paid to encourage removal of livestock from the market. In the case of dairy there is no change until 2005, and then a 15% cut in butter and milk powder intervention prices and an increase in quotas by 1.5% to be phased in from 2008 (CEC 1999a). Certainly, the six-year programme agreed at Berlin remains production-based, accounting for around 90% of the CAP budget. About 10% will go to Rural Development Programmes (RDP) under the Rural Development regulation (EC/1259/1999), which now covers specific support for organic farming, including the agri-environment measures. The total CAP expenditure has been fixed and should rise less than 10% over the six years, meaning that CAP expenditure as a percentage of the total EU budget has declined from 60% in 1989, to about 51% in 1994 and around 46% in 2001.

State support within the EU for the promotion of organic agriculture has largely existed under the CAP since 1992 or earlier for voluntary national measures, and to a lesser extent under structural, regional and research policies. There are four broad categories for the state support of organic agriculture: payments to producers, marketing and regional development, legal definitions (including standards) and information provision. Prior to the Agenda 2000 reforms, the main EU regulations that had sought to promote organic production were via state support for conversion to and continuing organic production. Such support was under the framework of the EU extensification programme (EEC/4115/88) and under the agri-environment regulation (EEC/2078/92). Funding for organic projects also existed under regional and rural development policies: EU Structural Funds under Objectives 1 and 5b, and EU LEADER programme funding. Advice, extension and information, and training and education, received funding under regulation 2078/92. Also, ten organic research projects had been funded up to and including the fourth EU Framework research programme (Lampkin *et al.* 1999).

Under regulation 2078/92 the EU met 50% of the cost of schemes, and 75% in Objective 1 regions, member states providing the remainder up to a fixed maximum. Within these parameters national level support for organic farming has varied. All member states have supported both conversion and maintenance of organic farms with the exception of the UK and France (which provides ongoing support in just three of its 22 Regions). However, the UK does figure amongst those states with the longest period for conversion payments of five years. Some countries have varied payments depending on the sector. Austria offered higher rates for horticulture and vegetable production than for arable (House of Lords 1999). In the UK, the original scheme (the Organic Aid Scheme) offered flat rates for all land types and the lowest support of all the member states. From 1999, the revised Organic Farming Scheme (OFS) increased payments under a greatly expanded budget (from £1 million in 1998–1999, to £11.35 million in 1999–2000). Nonetheless, the OFS ran out money within four months of its first year of operation, as it underestimated demand. Rates are variable depending upon the condition of the land and whether it falls under any existing payment schemes, such as the Arable Area Payments Scheme (AAPS) (House of Commons 2001). The high level of CAP payments to arable farmers provides a context for the profile of organic farming in the UK, which is heavily concentrated in the south and west of England and Wales and in Scotland. The predominant organically farmed land in 2000 was grassland (87%), mainly for dairy and livestock production, compared with arable and horticultural production (at just 13%) (House of Commons 2001). One consequence is an inadequate supply of organic feed for livestock, hence a further reliance on imports. Further concerns have been voiced regarding the inadequacy of support for horticultural production due to particular conversion problems in this sector (House of Commons 2001).

In 2000, organic farms covered nearly 3% of agricultural land in the EU – a total of 125 000 registered farms on 3.3 million hectares – an increase of approximately 20% over the previous year (Soil Association 2001a:). A comparison of the figures for organic and in-conversion land (as a percentage of overall agricultural land) in selected EU member states is presented in Table 13.1. The growth in market demand signalled from the late 1990s led several countries to introduce further specific programmes in support of organic farming, including France, The Netherlands, Denmark and Italy (Soil Association 2001a). The take-up of organic conversion has been further helped by the premiums on prices for organic foods. In the UK, the premium varies between sectors, reflecting, to a degree, the extent to which the sector is reliant upon imported produce as opposed to domestic production. There is a premium range of 50–60% for cereals (75% import compared to 25% domestic production in 1999–2000) and fruit and vegetables (85% import to 15% domestic production). The range is about 15–20% for meat (30% import to 70% domestic) and dairy (40% import to 60% domestic) (Soil Association 2001a; House of Commons 2001). The overall premium range in more mature national markets is between 10% and 30%. There may be scope for future premium reduction in the UK if increased domestic production replaces some imports (more likely to be long-term) and economies of scale are realised in processing and distribution systems. Also, some conventional food prices (e.g. milk) may recover from their current low. Some large retailers are seeking to use their dominant market position to reduce the premium for their customers (House of Commons 2001).

The Rural Development Programme (RDP), under which all official organic support now comes, requires that single integrated rural development plans be drawn up at the most appropriate geographical area. It combined what were nine separate EU measures including regulation 2078/92 and Structural Fund Objectives 1 and 5b, and farm and rural business training schemes. The aims of the RDP are to improve agricultural holdings; guarantee the safety and quality of foodstuffs; ensure fair and stable incomes for farmers; and ensure that environmental issues are taken into

Table 13.1 The area of organic and in-conversion land for selected countries in Europe for 1999–2000 (The Soil Association 2001c).

Country	1999–2000 (ha)	percentage agricultural land (ha)
Sweden	268 000	11.2%
Austria	345 000	10%
Denmark	160 000	6%
Germany	422 000	3%
Italy	900 000	5.3%
UK	425 000	2.3%
France	220 000	1%
EU average		2.2%

account. Also, the regulation aims to develop complementary and alternative activities that generate employment, with a view to slowing the depopulation of the countryside and strengthening the economic and social fabric of rural areas; and to improve living and working conditions and equal opportunities (CEC 1999b).

Nic Lampkin and his co-authors (1999) viewed the initial RDP proposals favourably as they had the potential to integrate the four broad categories that they identified as promoting organic production. However, the reform does not generate any new funds from the CAP. Rather, the Berlin agreement passed the initiative back to the member states. It was agreed that member states could reduce conventional CAP compensation payments to farm businesses by up to 20% (termed 'modulation'), and then, by adding matched national funds, could redirect these funds to other measures such as the agri-environment schemes (Ward & Falconer 1999; Falconer & Ward 2000). Amongst the countries that have adopted modulation the implementation has varied. For example, the UK chose to transfer 2.5% (rising ultimately to 4.5% in 2005–2006) of payments to other measures at a flat (not progressive) rate, supporting the *status quo* of payments favouring larger, usually arable, farmers. Conversely, France introduced a degree of progressive modulation redistributing some funds to smaller farmers. The total package within the UK, for instance, during the period 2001–2006, is worth £1.6 billion which is composed of EU receipts, UK CAP and UK matching funds. There is £1 billion for agri-environment measures of which there is £139 million for payments to convert to organic farming. Such payments are an improvement on past UK support, but will do little to meet the forecasted market growth in demand for organic food. The agri-environment spending is increased but the extent of modulation falls well short of the 20% possible, lessening the potential benefits of the reform (Falconer & Ward 2000).

A further strategy for the UK was the introduction of an Organic Targets Bill to the British Parliament, which called for an organic conversion target of 30% of UK agricultural land by 2010. The proposal was a symbolic initiative lacking the government support necessary to make it law, despite garnering widespread cross-party support amongst members of parliament. Nonetheless, the government has signalled a desire to engage in a debate about the future of farming and food production in the UK following the most recent foot-and-mouth outbreak (Labour Party 2001). The Soil

Association has called for the introduction of an organic stewardship payment (explicitly recognising the public goods role of organic farming) in addition to conversion payments that are merely compensation for the initial period when farmer earnings fall (House of Commons 2001). Regional water authorities have recognised the contribution that organic farming can make to reduction of water pollution from pesticides and nitrates. Wessex Water have offered a further £40 per hectare subsidy over two years to farms undertaking organic conversion (House of Commons 2001).

The increasing spread of organic farming concurrent with the introduction of large-scale field trials for genetically modified (GM) crops has wrought further conflicts over the regulation of the agri-environment. Disagreements exist over specific management issues such as the separation distances necessary to prevent cross-pollination of non-GM with GM crops, while the biotechnology industry questions the validity of the organic farming and food sector adopting a non-GM or GM-free stance for its products (Barling 2000). The co-existence of organic farming within the conventional system model is made all the more difficult by such conflict. The introduction of EU regulation for organic seed standards, due for 2003, provides a further context for the conflict between organic farming and the proponents of GM crops to be played out.

The future promotion of organic farming under the CAP is linked to a number of considerations. In one respect it plays a part in the debate of the future of food and farming in the EU, as it provides a clear alternative system. On the other hand, the advance of organic farming to date has been promoted within the conventional system. Hence, the expansion of organic farming within the CAP is linked to the further promotion of the RDPs, conceptualised as part of a second pillar of the CAP seeking an integrated rural policy. Within this concept, organic farming can be seen both as an extensification strategy and the provider of wider public goods, notably in terms of environmental protection, biodiversity enhancement, nature and cultural landscape conservation and rural development. Organic farmers would then gain in terms of compensatory payments for foregoing other profits (from intensive production) in order to promote these public goods. This concept of public goods fits within the broader multifunctional model of agriculture developed by the EU and presented as part of its negotiating position within the revision of the AoA (see below). However, to what extent will the CAP move further from its productionist impulse and move resources to the second pillar?

The Agenda 2000 reform has built in a mid-term review for 2003, aimed at the main sector regimes. The present Agriculture Commissioner, Fischler, is seeking to widen the review to accelerate the move from production support to extensification and the second pillar. He has questioned, for example, the legitimacy of 45% of the CAP's budget going on arable payments (Agra Europe 2001). Such change is contingent upon the intergovernmental politics of the CAP policymaking process, and therefore upon the national political configurations informed by fifteen different national farming contexts up to 2003. In 2001 there were some positive signs of change at the national level, including an environmental reform-minded group of agriculture ministers in Sweden, Denmark and Austria and somewhat mixed messages from the UK. In Germany the Green Party's Renate Kunast took over a reformed ministry, now called

Consumer Protection, Food and Agriculture, and set a national target of 20% organic production by 2010 (Kunast 2001). This suggests that there might be a shift in the position of the Franco-German alliance that is at the core of the CAP process. In the meantime, elections loom in Germany and France in 2002. The French election will restrict the scope for radical redirection of payments in the short term, and a change of government in Germany might impede wider reform efforts (Grant 2001). In addition, the scope for further change will be played out against the bigger canvas of the AoA and its review, to which we can now turn.

The Agreement on Agriculture and related WTO agreements: trade liberalisation, decoupling and organic farming

The AoA marked a step towards greater trade liberalisation in agricultural products: an area marked by protectionist policies in the recent past. However, free trade in agriculture remains an ideal rather than a reality. The OECD reported that market and trade distorting support to agricultural production in 1999, at 40% of gross farm receipts, was back to levels of the mid-1980s (as measured by the Producer Subsidy Estimates that included all forms of state support) (OECD 2001). To date, studies evaluating the impact, or potential impact, of the liberalisation agreements on agriculture have drawn varying conclusions. Some initial studies suggested that trade liberalisation would be beneficial to agri-environmental farming, as the fall in price support would reduce production and the amount of agrochemicals applied to the land. Further study in the UK concluded that the net environmental impact would be damaging, particularly in more marginal farming areas (Potter *et al.* 1999). Alan Swinbank has suggested that UK agriculture would survive and be competitive in a more liberal world market, but the social cost would be great with more small farms disappearing and further concentration (House of Commons 2000a). In less developed countries increased agricultural trade liberalisation has led to an increase in food imports which weakens food security (FAO 1999), while the World Bank reported that the overall effects of trade liberalisation on the world's poorest populations have been adverse (Lundberg & Milanovic 2000).

The AoA adopted a three-pillar approach to agricultural trade liberalisation focusing on the three main supports for farming (as illustrated in the CAP): export competition, market access for imports and domestic support. In order to facilitate fairer export competition, the agreement sought to reduce export subsidies whereby financial incentives are given to export products. The EU was the main user of such direct export subsidies. The reductions were negotiated against a baseline of 1986–1988 levels when the EU's export subsidies were at an all-time high, thus lessening the impact of this reform (Grant 1997; Einarsson 2000). Other forms of state-led (indirect) export support, such as export credits, production-linked food aid and state trading enterprises, were also excluded from the agreement, protecting such practices amongst other food exporters such as the US and New Zealand. A variety of protectionist methods such as import levies and quotas restrict or deny market access. Under the AoA all forms of protection have to be merged into fixed tariffs (a process called tariffication). Tariffs were to be reduced by allocated amounts with a minimum level of market access to be allowed, varying according to specific agricultural product. At the centre of the AoA, however, were the arrangements for the reduction of domestic support.

The different types of domestic support were subject to different arrangements under the agreement. These differing arrangements are known as boxes and colour-coded. Into the red box went supports that were now banned. The amber box contained production-related price supports, subsidies and payments. The total support of this kind, deemed to be most distorting to trade, is calculated as the Aggregate Measure of Support (AMS). Within the AMS production supports, such as intervention buying, had to be reduced by a certain percentage. This differed for developed countries (by 20% over 6 years), developing countries (13.3 % over 10 years), and least developed countries (protected from reductions under special and differential treatment under article 15). There were further details to the amber box, but overall it did signal a decrease in agricultural production price supports. However, some domestic supports were exempted from reductions (but included in the AMS calculation) and put into a blue box. The blue box, a compromise result of tough negotiations at the end of the round, included exemptions for direct payments under programmes that sought to limit production (Wolfe 1998; Swinbank 2000b). Foremost amongst these exemptions were the CAP's direct payments to arable farmers (AAPS) and livestock farmers, introduced under the McSharry reforms and continued under Agenda 2000 (and listed in Article 6). Hence, under the CAP reforms such payments were seen as AoA-compliant, at least for the present.

Lastly, the green box contained domestic supports that were exempted from reduction commitments because they met 'the fundamental requirement that they have no, or at most minimal, trade-distorting effects or effects on production' (Annex 2). Also, they did not have 'the effect of providing price support to producers' (Annex 2). Such supports were viewed as being truly decoupled from production. Examples of such payments included:

- research
- domestic food aid and food security measures
- direct income to producers where it is provided as decoupled income support
- payments under environmental programmes, and
- retirement of producers or resource retirement (land removed from production).

The conflicting negotiating positions resulted in significant compromise, producing ambiguous wording in the final agreement (Wolfe 1998; Swinbank 2000b). Consequently, the agreement included a review process that started in 2000. There is a peace clause for the EU until 2003, after which it is unclear if some elements of the agreement (those in the blue and green boxes) will be open to challenge as trade distorting under the general GATT/WTO rules for trade disputes (Swinbank 2000b). The review of the agreement will hold the key to the future viability of many nations' domestic support arrangements. Several countries have included elimination of the blue box payments in their official negotiating positions for the review, posing a potential problem for the EU. One way out of this problem would be to decouple the direct payments from production and move them into the green box. This would mean significant changes to the Agenda 2000 arrangements, and a significant shifting of resources to the RDP and the second pillar.

The EU's negotiating position is to maintain the blue box payments, while promoting the multifunctional role of agriculture. The multifunctional role 'covers the protection of the environment, and the sustained vitality of rural communities, food safety and other consumer concerns' (WTO 2000). The multifunctional model evolved from the so-called European model of agriculture and emerged in 1997 from within the CAP establishment, promoted by the Commission and the COPA (the entrenched Euro-farmers organisation). This conservative approach to multi-functionality was greeted with some scepticism, being regarded as thinly veiled protectionism. The EU's linking of the maintenance of blue box payments to the multifunctional model has reinforced this scepticism to some extent. Nonetheless, a more progressive version of the concept, applied to decoupled payments eligible for the green box, has gained a degree of favour with agricultural economists (Guyomard *et al.* 2000; Anderson & Morris 2000; Swinbank 2000b; Tangermann 2001). This version has been promoted by environmental and organic organisations (Crompton & Hardstaff 2001; Azeez 1999; House of Commons 2000b).

The movement of the CAP payments to the green box would signal a move to more extensive and environmentally friendly farming methods and would be very beneficial to the development of organic farming. There are two possible obstacles to this. Firstly, while the AoA allows domestic support for decoupled non-trade concerns (such as environmental protection), there remains the caveat that such measures must have 'no, or at most minimal, trade distorting effects or effects on production'. English Nature (a public sector conservation agency) has expressed concerns that this clause could be used to prevent some environmental payment supports for farming, as they may have some impact upon production levels and so on trade (House of Commons 2000b). Agricultural economists have also highlighted this possibility (Swinbank 2000b; Tangermann 2001). Hence, there is a perceived need for the review of the AoA to provide a clear and final definition of what constitutes legitimate decoupled payments (i.e. non- or minimally distorting): a decision which will impact upon organic farming and its claims for payments for its role as a provider of public goods. The negotiating position of the Cairns Group (made up of the major agricultural commodity exporting nations, apart from the US and EU), led by Canada, has called for a review of the green box to ensure a clear understanding that such programmes are at the most minimally trade distorting (Crompton & Hardstaff 2001). The outcome of the review could have, therefore, important ramifications for more extensive farming. The second major obstacle, of course, remains the will of the member states to reform the CAP further towards the second pillar supports. The signs are distinctly mixed, as indicated previously.

A final issue affected by the trade rules that concerns the organic movement is animal welfare. Under the rules of the GATT, the AoA and the Technical Barriers to Trade (TBT) agreement, distinctions made between products have to be based on the characteristics of the final product. Discrimination between 'like products' is not permitted. Distinctions cannot be based on process and production methods (PPMs) that would specify animal welfare standards. A free-range egg is the same as a battery hen's egg. Within the EU there has been a drive to raise animal welfare standards, including for farmed animals. Within the trade rules the perceived key problem is one of competitiveness (Fisher 2000). The costs of EU standards will allow international or third country producers to undercut the European producers. Consequently, the EU's directive on the welfare of laying hens, that seeks to phase out the use of battery cages by 2012, has a review in 2005 that has to take into account developments in WTO negotiations (House of Commons 2000a). Animal welfare organisations are seeking to get changes in the WTO's rules, or at least a revised clarity, starting with the review of the AoA (RSPCA 1999).

The EU has suggested three possible strategies for supporting animal welfare standards; firstly, negotiating international multilateral agreements on animal welfare, aside from the WTO arrangements, at best a long-term strategy; secondly, introducing either a voluntary or mandatory labelling scheme. It has been argued that the TBT does not explicitly prohibit a mandatory labelling scheme based on PPM, although trading partners would have to agree with such a scheme, which may be unlikely to happen. Also, welfare labelling would help with primary food products but be more difficult with processed and catered products (Fisher 2000). In the negotiating position for the review, the EU put forward the third strategy, seeking to add animal welfare support to the green box measures (WTO 2000). The EU has yet to support this strategy with payment supports under the CAP (Fisher 2000).

A more radical reform of the AoA has also been advocated. Tim Lang and Colin Hines have argued that the GATT should become a GAST, General Agreement for Sustainable Trade (Lang & Hines 1993). A reworking of first principles of the AoA could emphasise the sustainability impacts (Crompton & Hardstaff 2001). Emphasis on rules that favour local food economies, healthy food production and animal welfare would be of benefit to organic farming. These principles would allow for the more holistic approach of organic farming to take centre stage. As things stand, the incremental moves to greater extensification of production and decoupled payments for recognised and valued public goods generated through appropriate farming methods are the more likely, but still far from certain, avenues for reform of the world trade rules.

Conclusions

The reform of the CAP and the revision of the AoA will be played out in concert with debates at the national level across Europe on the future shape of agriculture and food production. The outcome of these debates will be decided within a multilevel framework of governance for food, at the local, national, regional and global levels. Decisions taken at these interrelated levels will shape the more individual, local-based decisions of farmers over whether to take up (or continue) more extensive and sustainable agricultural practices such as organic farming. Significant increase of state support for extensive farming systems, at the expense of production supports, will undoubtedly have an impact on the nature and spread of farming and its landscape in countries such as the UK. Progress towards policy agreements that are more conducive to the support of organic farming will also help the regulators of farming and food to catch up with the surge in market demand for organic produce in the EU. With the growth of the organic food market we may witness further change to the structures of the organic food industries themselves.

The international regimes of the CAP and the WTO's AoA do not provide settings particularly conducive to the promotion of a holistic alternative to the conventional model of farming. At national level, in the UK at least, organic farming has paid a price for entry into the dominant agricultural policy paradigm. Organic farming has downplayed its role as an alternative system, in favour of promoting itself as offering strategies for extensification of farming systems and sympathetic approaches to the management of the agri-environment and its biodiversity. Similarly, within the CAP the expansion of support for organic farming has been linked to extensification and environmental needs as the EU has struggled to shift significantly its resources from the dominant productionist support to a more integrated rural development policy. This largely intergovernmental struggle still provides a crux to further reform, as fifteen different member states seek to reconcile the needs of their agricultural and processing producer groups with the redirection of resources under the CAP. The promotion of the public goods role of less intensive farming has evolved within the rural development agenda. The formulation of the multifunctional model of agriculture by the EU has also encompassed the public goods role, although the vision has been muddied somewhat by the continuing advocacy of certain production-linked direct payments within this model. The decoupling of such supports from production does offer a way forward for support for organic farming under the international trade rules. There remain, however, much work and clarification to be undertaken within the terms of the review of the AoA for such progress to be realised.

References

Aaarnik, W., Bunning, S., Collette, L. & Mulvany P. (1999) *Sustaining agricultural biodiversity and agro-ecosystems functions.* Report of an international technical workshop organized jointly by FAO and the secretariat of the Convention on Biological Diversity with the support of the Government of The Netherlands. Food and Agriculture Organisation, Rome.

Agra Europe (2001) Fischler raises 'morality' of EU arable aid. *Agra Europe,* 1936, 26 January, 7.

Altieri, M. (1995) *Agroecology: The Science of Sustainable Agriculture.* 2nd edn. Westview Press, Colorado.

Anderson, K. & Morris, P. (2000) The elusive goal of agricultural trade reform. *Cato Journal,* 19 (3), 385–96.

Azeez, G. (1999) Personal communication with Gundula Azeez, policy officer for The Soil Association, October 1999.

Azeez, G. (2000) *The Biodiversity Benefits of Organic Farming.* The Soil Association, Bristol.

Barling, D. (2000) GM crops, biodiversity and the European agri-environment: Regulatory lacunae and revision. *European Environment,* 10 (4), 167–77.

Beus, C. & Dunlap, R. (1990) Conventional versus alternative agriculture: The paradigmatic roots of the debate. *Rural Sociology*, 55, 590–616.

Buck, D., Getz, C. & Guthman, J. (1997) From farm to table: The organic vegetable commodity chain of Northern California. *Sociologia Ruralis*, 37 (1), 3–20.

CEC [Commission of the European Communities] (1958) *Treaty of Rome*. CEC, Brussels.

CEC [Commission of the European Communities] (1991) Council Regulation (EEC) 2092/91 on organic production of agricultural products and indications referring thereto on agricultural products and foodstuffs. *Official Journal*, L 198, 22 July.

CEC [Commission of the European Communities] (1994) *Organic farming: Green Europe 2/94*. Office for Official Publications of the European Communities, Luxembourg.

CEC [Commission of the European Communities] (1996) *Progress report from the Commission on the implementation of the European Community programme of policy and action in relation to the environment and sustainable development 'Towards Sustainability'*. Com (95) 624 final. CEC, Brussels.

CEC [Commission of the European Communities] (1999a) *Berlin European Council: Agenda 2000, Conclusions of the Presidency*. European Commission Directorate-General of Agriculture Newsletter, 10 March.

CEC [Commission of the European Communities] (1999b) Council Regulation (EC) 1259/1999 establishing common rules for direct support schemes under the common agricultural policy, *Official Journal* L 160, 0113–8.

Clunies-Ross, T. & Cox, G. (1994) Challenging the productivist paradigm: Organic farming and the politics of agricultural change. In: *Regulating agriculture*, (eds P. Lowe, T. Marsden, & S. Whatmore), pp. 53–73. David Fulton Publishers, London.

Council of Ministers (2001) 2360th Council meeting – Agriculture: Press Release, 9930/99301 (241), 19 June, Luxembourg.

Crompton, T. & Hardstaff, P. (2001) *Eat This: Fresh Ideas on the WTO Agreement on Agriculture*. RSPB, Sandy, UK.

Einarsson, P. (2000) *Agricultural trade policy as if food security and ecological sustainability mattered: Review and analysis of alternative proposals for the renegotiation of the WTO Agreement on Agriculture*. A report from Church of Sweden Aid, Forum Syd, Swedish Society for Nature Conservation and Programme of Global Studies. Forum Syd, Stockholm.

Falconer, K. & Ward, N. (2000) Using modulation to green the cap: the UK case. *Land Use Policy*, 17 (4), 269–77.

FAO [United Nations Food and Agriculture Organization] (1999) Synthesis of Case Studies, X3065/E. *FAO symposium on Agriculture, Trade and Food Security*, Geneva, 23–24 September 1999. FAO, Rome.

Fisher, C. (2000) EC Consultation meeting on agriculture: Discussion paper presented on behalf of the Eurogroup for Animal Welfare. Brussels, 6 November 2000.

FSA [Food Standards Agency] (2000) Position paper: Food Standards Agency view on organic foods. FSA, London, 23 August 2000.

Goodman, D., Sorj, B. & Wilkinson, J. (1987) *From Farming to Biotechnology: a Theory of Agro-Industrial Development*. Blackwell, Oxford.

Grant, W. (1997) *The Politics of the Common Agricultural Policy*. Macmillan, London.

Grant, W. (2001) Is real reform now possible? (http://members.tripod.com/~WynGrant/Overview.html accessed 22 May 2001.)

Guthman, J. (1998) Regulating meaning, appropriating nature: The codification of California organic agriculture. *Antipode*, 30 (2), 135–54.

Guyomard, H., Bureau, J-C., Gohin, A. & Le Mouel, C. (2000) Impact of the 1996 US FAIR Act on the Common Agricultural Policy in the World Trade Organisation context: the decoupling issue. *Food Policy*, 25, 17–34.

House of Commons (2000a) *The implications for UK agriculture and EU agricultural policy of trade liberalisation and the WTO round.* House of Commons Agriculture Committee Sixth Report, Vol. I, HC 246-I. HMSO, London.

House of Commons (2000b) *The implications for UK agriculture and EU agricultural policy of trade liberalisation and the WTO round.* House of Commons Agriculture Committee Sixth Report, Vol. II, Evidence and appendices, HC 246-II. HMSO, London.

House of Commons (2001) *Organic farming.* House of Commons Agriculture Committee Second Report, Vol 1, HC 149-I. HMSO, London.

House of Lords (1999) *Organic farming and the European Union: House of Lords Select Committee on the European Communities Sixteenth Report.* Vol. I, HL 93. HMSO, London.

IEEP [Institute for European Environmental Policy] & WWF [World Wide Fund for Nature] (1994) *The Nature of Farming. Low Intensity Systems in Nine European Countries.* IEEP/ WWF, London and Geneva.

Kunast, R. (2001) The Magic Hexagon. *The Ecologist*, 31 (3), 48–9.

Labour Party (2001) *Ambition for Britain: Labour's manifesto 2001.* Labour Party, London.

Lampkin, N., Foster, C., Padel, S. & Midmore, P. (1999) *The Policy and Regulatory Environment for Organic Farming in Europe: Synthesis of Results. Organic Farming in Europe: Economics and Policy, Volume 1.* Universitat Hohenheim, Stuttgart.

Lang, T. (1999) The complexities of globalization: The UK as a case study of tensions within the food system and the challenge to food policy. *Agriculture and Human Values*, 16, 169–85.

Lang, T. & Hines, C. (1993) *The New Protectionism: Protecting the Future Against Free Trade.* Earthscan, London.

Lang, T., Barling, D. & Caraher, M. (2001) Food, social policy and the environment: Towards a new model. *Social Policy and Administration*, 35 (5), 538–58.

Lobstein, T. (1999) Organic standards: Where do we draw the limits? *Food Magazine*, 44, 14–15.

Lundberg, M. & Milanovic, B. (2000) The truth about global inequality *Financial Times*, 25 February 2000.

Mander, J. & Goldsmith, E. (1997) *The Case Against the Global Economy.* Sierra Club Books, San Francisco.

Meikle, J. (2000) Meacher praises organic farming. *The Guardian*, 26 May 2000.

OECD [Organisation for Economic Co-operation and Development] (2001) *Agricultural Policies in OECD Countries: Monitoring and Evaluation 2000.* OECD, Paris.

Pinstrup-Andersen, P., Pandaya-Lorch, R. & Rosegrant, M. (1999) *World food prospects: Critical issues for the early twenty-first century. 2020 Vision Food Policy Report.* International Food Policy Research Institute, Washington, DC.

Podbury, T. (2000) US and EU agricultural support: Who does it benefit? Australian Bureau of Agricultural and Resource Economics. *Current Issues*, 2, October.

Potter, C., Lobley, M. & Bull, R. (1999) *Agricultural liberalisation and its environmental effects.* Report commissioned by Countryside Commission, Countryside Council for Wales, English Nature and Scottish Natural Heritage. Wye College University of London, Wye.

Pretty, J.N. (1998) *The Living Land.* Earthscan, London.

Pretty J.N., Brett, C., Gee, D. *et al.* (2000) An assessment of the total external costs of UK agriculture. *Agricultural Systems*, 65, 113–36.

RSPCA [Royal Society for the Prevention of Cruelty to Animals] (1999) *WTO: Food for Thought.* RSPCA Campaigns, Horsham.

Soil Association (2001a) *The Organic Food and Farming Report 2000.* The Soil Association, Bristol.

Soil Association (2001b) *Organic Farming, Food Quality and Human Health: a Review of the Evidence.* The Soil Association, Bristol.

Soil Association (2001c) *Fact Sheet: Organic Facts and Figures, August 2001.* Soil Association, Bristol.

Swinbank, A. (2000a) EU Agriculture, Agenda 2000 and the WTO Commitments. *World Economy*, 22 (1) 41–54.

Swinbank, A. (2000b) Memorandum submitted to the House of Commons Select Committee in Agriculture inquiry on the implications for UK Agriculture and EU agricultural policy of trade liberalisation and the WTO round. 14 January.

Tangermann, S. (2001) Common Agricultural Policy facing a twin challenge of enlargement and the next WTO round. Programme abstracts, *Food Chain 2001 Conference*: Uppsala, 14–16 May 2001.

Thomas, C. (1999) Introduction. In: *Global trade and global social issues*, (eds A. Taylor & C. Thomas), pp 1–13. Routledge, London.

Watts, M.J. & Goodman, D. (1997) Agrarian questions: Global appetite, local metabolism. Nature, culture and industry in *fin-de-siècle* agro-food systems. In: *Globalising food: Agrarian questions and global restructuring* (eds D. Goodman & M.J. Watts), pp. 1–34. Routledge, London.

Ward, N. & Falconer, K. (1999) Greening the CAP through modulation: opportunities and constraints. *Ecos*, 20 (2) 43–8.

Whitby, M. (1996) Losers and gainers in rural policy. In: *The rural economy and the British countryside* (eds P. Allanson & M. Whitby), pp. 167–86. Earthscan, London.

Wolfe, R. (1998) *Farm Wars: the Political Economy of Agriculture and the International Trade Regime.* Macmillan, London.

WTO [World Trade Organisation] (2000) *EC comprehensive negotiating proposal*, Committee on Agriculture Special Session, WTO, G/AG/NG/W/90, 14 December.

Chapter 14

Integration

Learning to Solve the Pesticide Problem

Peter Groenewegen, Frank den Hond and Nico M. van Straalen

Introduction

The central question of this volume is: why is the pesticide problem so persistent? Our assumption is that the persistence of problems related to pesticide use is to be found in the origin and structure of pesticide use. In this concluding chapter we aim at integrating the findings from the preceding chapters into an overall framework.

The starting point of our effort was a widely found assertion that can be summarised in two sentences:

(1) The use of crop protection technology in agriculture has brought enormous benefits but it also generates negative impacts on the environment.
(2) Current food supply and agricultural interests are at odds with the perspective of sustainability.

These juxtapositions can be regarded as a starting point for a discussion on solutions. However, they lend themselves easily to black-and-white schematising, which we consider unproductive. Too often, the emphasis in the discussion is on moral or ethical aspects of the arguments used, on the 'goods' and 'bads' of benefits and impacts. In turn, this emphasis blocks a clear view on the arguments themselves, for example their robustness against empirical evidence, or in some instances even the quality of their empirical foundation, if there is empirical evidence at all. The result of such a debate may quickly become to resemble a situation that has been called an 'artificial controversy' (van Dommelen 1999), a stalemate from which there is no easy escape and to which there are certainly no simple solutions. In the introduction to this volume we tried to set out a scheme to analyse the juxtapositions by going into more detail with regard to a number of underlying spheres of interaction. We distinguished the following interacting spheres as a logical inroad: agricultural innovation, agricultural production and socio-economic institutions. Subsequently, on the basis of both disciplinary and interdisciplinary insights, each of the chapters in this volume went into some detail regarding problem analysis, and discussed barriers and solutions to the origins and causes of the persistence of the pesticide problem, albeit in different balances between problem analysis and discussion of barriers and solutions.

In this final discussion we want to interrelate the contributions made in each of the preceding chapters in more detail. We can consequently provide a somewhat more complete understanding of the dynamics that occur within and across our spheres. While the three spheres are highly interrelated, we distinguished between them in the introductory chapter primarily as an analytical device. The constituting configurations of actor groups within each of the spheres led to different relevant factors and dynamics that suggest that each of these spheres can be broken down into more or less consistent sub-systems that are only loosely coupled to each other. For instance, the authorisation system is part of the socio-economic sphere. It has been designed to avoid harmful consequences for the environment by the *ex ante* evaluation of chemicals. However, the dynamics within the authorisation system are bound into a system of laws and rules that govern the interactions of relevant parties – such as authorisation offices, test laboratories, and agrochemical companies – and grows largely on its own accord, forcing companies to submit increasingly costly evidence. Moreover, the suggestion that state-of-the-art science is used is in part a legal fiction. Redressing the balance towards a more dynamic system will be hard but is essential to a more dynamic and innovative system. The EU harmonisation effort shows that the time taken to encompass change should be counted in decades rather than years. Analysis of such configurations provides insight into the solution of agricultural and environmental problems of pesticides.

Breakdown of the problem and future options

The dilemmas of regulating and guiding the development of pesticide technologies and application practices that may affect the environment are documented in a variety of ways in this volume. The simple representation of a system that we have drawn in the introductory chapter can be sharpened along two different lines of reasoning. The first is to use the analysis for the identification of barriers and solutions within the boundaries of each of the spheres, as well as the possibility to modify existing interactions with the other spheres. For instance, in order to improve the existing regulatory system it could be advisable to optimise decision making processes and invest in the relevant methods to support improved decisions. Or similarly, at the farm level – in order to change within the system towards technologies that minimise environmental impacts – to make sure that barriers are made properly and the incentive scheme is working in the right manner. The second is to identify the potential for transformations of the practices towards other processes and interactions that avoid basic design faults in the longer term. However, such a multi-levelled approach would require subtle analysis and consideration of arguments in order not to misuse information and conclusions from the chapters. Therefore, we discuss these possibilities primarily from the first line of reasoning within each of the spheres. The framing of conclusions in a conceptual framework of transformation, such as the second line of reasoning would require, is only undertaken where we have appropriate information.

Agricultural production

The basic questions with regard to pesticide problems are connected to the application of pesticides at farm level. The factors that form an explanation for the current pattern of adverse effects can be broken down into different sets. The actual problems that are discussed focus on optimisation of decision making at the farm level from the perspective of environmental benefits without directly challenging the existing role of chemicals in the agricultural production system. Optimisation can be regarded from the perspective of economic utilisation of available technical options, and that of the quality of agricultural practice. Both perspectives may inform the development of a proper incentive structure, based on the design of regulatory and other measures, that can make farmer choices shift.

The use of technology in agricultural production is not optimal, nor is it straightforwardly based on rational decision making. As Wossink & de Koeijer (Chapter 6) have argued, the use of technology certainly does not satisfy an optimal choice of technical options with regard to environmental performance. Considerable room for improvement in this area exists. Persistent differences in environmental efficiency between farmers were demonstrated. A crucial factor that may influence decision making is lack of information. Additional information that is used to help to estimate the economic and environmental effects improves such farmer decisions, but optimal decision making at the farm level is not ensured by education alone. A 'right' combination of regulatory and economic incentives will be necessary to assure improved balance in such decision making. Farmers are not only risk-averse profit seekers who operate under conditions of bounded rationality. However, what they are beyond this characterisation remains an open question, and, moreover, may differ across a population of farmers. Consequently, what is the 'right' combination might be contingent upon a range of different factors.

De Snoo (Chapter 7) analyses a different approach to the adjustment of agricultural production. Although his basic arguments are similar to an economic approach, the focal point is the optimisation of environmental performance. The variation in environmental impact between various pesticide application methods can be considerable. A strategy directed at the reduction of environmental impact may consist of a combination of methods including the choice of pesticide products, the implementation of buffer zones, and technical adjustments in mode of application. In addition, abatement measures directed at the worst polluters may have significant overall benefits. A one-third reduction in damage may be reached when using this approach. With regard to implementation, the author suggests that voluntary or market-based regulation of such certification schemes may be the preferred option. It might be inefficient for regulators to identify unequivocally which farmers are the 'worst polluters'.

There is a linkage between the actual agricultural production methods used and the policy that can be designed. As argued by Parris & Yokoi (Chapter 8) there is still a quite urgent need to improve the statistics with regard to the use of pesticides in OECD countries. Adequate policies require available data, not only overall pesticide use data, rather than sales data, or data expressed in kilograms of active ingredient, but broken down to the actual use of pesticide products (including frequency of use and application techniques used) at farm or even field level. Also, what is needed are models that are developed on the borderline between science and policy and which enable the assessment of risks in practice. Such additional indicators are deemed necessary to assess the effectiveness of policy. This holds with regard to the variety of policy measures that may directly address imbalances in the regulatory system, the policy mix directed at farmers as well as the need to assess combination and long-term effects of actual pesticide use. The work they describe is in progress and based on the integration of improved sales and/or use figures and information on pesticide hazard and exposure. A more careful selection of the information that is required and calculation of risks in a more precise manner also may help to inform policy with regard to the most effective approaches to the management of the risks of pesticide usage.

The combination of these three chapters provides interesting perspectives and further questions. According to de Snoo, part of the problem is sloppy farming, disinterest among farmers, and lack of incentives to do better; witness the observed difference in overuse between costly herbicides and cheap insecticides. An easy opportunity would be to increase the price of the most polluting pesticide products – environmental economists would tell you to do precisely this. However, two problems arise. Along the lines of Wossink & de Koeijer, one may question the efficacy of such measures. Farmers' responses to such price increases are not likely to be fully rational, if there are no other accompanying incentives that help them change their pesticide use behaviour. Such a price increase is supposed to invoke innovation in pesticide use, but the diffusion of such innovations is not efficient. The other problem relates to Parris & Yokio's argument. To the extent that a shift in relative prices of pesticide products results in shifting farmer preferences for specific products over other products, it depends on exactly how this shift materialises whether there will be beneficial or detrimental effects on the overall environmental impact.

In answer to our central questions, there is an increased awareness that the vision on agricultural production is in need of change. Agricultural production is to be viewed as an internally differentiated system. Increased attention needs to be paid to the micro-management of agricultural production at farm or even field level. A better understanding of actual farmer behaviour and competences in making the 'right' decisions on pesticide use in agricultural production should have consequences for agricultural policy and the policy instruments that need to be deployed. Two examples come to mind. One example is the general trend of cutting on public spending on farmer education and extension services, which is to be criticised. These functions are important in, and should be oriented at, sensitising farmers *in their specific situation* to the various ways in which the production factors at hand can be deployed, as well as the interdependencies between them, and their external effects. It remains to be seen whether the current privatisation of these functions will bring about a system of farmer education and extension services that is sufficiently 'tailor-made' in this respect. Another relates to the concentration of government support for new technologies in agricultural production. Much of such support is decontextualised in the sense that the research directions are grafted on an ideal-type conception of agricultural production ('technology push' instead of 'market pull'). Moreover, it can also be argued that the implementation of such technology as currently taking place is far from optimal; it is left to the forces in the marketplace. It would appear, then, that public spending on improving the quantity and above all the quality of agricultural production could be much more effective by adopting a practice-driven view of agricultural production.

Agricultural innovation

The high dependency on chemicals to manage pests has led to a number of distortions that are in part a consequence of the dynamics of the use of synthetic pesticides in agricultural production itself. As Struik & Kropff (Chapter 2) argue, there is growing awareness that the chemical era of crop protection has to come to an end. One of the elements in agricultural production that needs to be changed is the emphasis in research. While substantial effort (and funding) has been spent on chemical solutions to specific pests, substantially less effort has been made to develop a prevention system based on agro-ecological insight. Struik and Kropff distinguish between three different approaches that can be used to structure solutions for the reduction of chemical dependence:

- refinement of the application of technology, i.e. innovation in technical applications (*cf.* discussion in Chapter 10)
- an increased emphasis on the development of agricultural practices based on integration of a broad range of preventive and non-chemical control measures (Chapters 6 and 7)
- the consequent implementation of non-chemical routes (Chapters 12 and 13).

Within mainstream farming until recently the application of (chemical) technology was thought to provide a systemic input that improves agricultural productivity significantly. Therefore attention to the technology push forces within the agricultural system is warranted. Some important lessons are to be drawn from the material presented in this volume. One of the first is that technological development is still rather piecemeal and little directed at the improvement of the overall system (*cf.* Carr, Chapter 11, on Bt corn). The positioning of new chemicals for profitable mass markets drives innovation in the agrochemical industry. The search and innovation logic within the industry tends to be reinforced by structural characteristics of the existing agrochemical companies, their markets and search of profitability (den Hond, Chapter 4). Some of their logic is also directly connected to the reasoning and structures of the regulatory environment (Irwin & Rothstein, Chapter 5). These structural features inhibit in part the search for alternative solutions that could be supported by the industry. The combination of forces of intense agriculture and high-cost research does not seem easy to breach. Thus, there is not a direct impetus within the technology-based industry to support alternatives in agricultural production.

Nevertheless, the innovation emphasis may also be shifted from the innovation in technologies to the innovation in production practices. With regard to this 'softer' side of agricultural production there is more room for improvement (Chapters 2, 6, and 7). There may also be the opportunity to make combinations from this perspective that suggest the development of new technological support tools. One example is the fine-tuning of the control of weeds and possibly other pests that becomes possible with information and sensory technology: the development of site-specific pest management (Swinton, Chapter 10). According to Swinton, the further development of these technologies is dependent on the availability of a diverse base of pesticides, as well as the relevant information technology. The resulting positive consequences of such a strategy relate to a reduced area of exposure, not necessarily a reduction in dose, which is currently the mainstream approach to reduction of environmental damages. However, some of the pesticide products that would be required may be those that pose serious problems today. The rate of development of new pesticide products may slow down if there are limited incentives for the industrial development of new active ingredients, if limited markets reduce the prospect of recovering the expenses of R&D costs. Consequently, or paradoxically, such an approach relies on a technology control strategy that already today is difficult to realise, as was exemplified in the preceding section. In the solution to this conundrum a combination of technology and government rules seems to be most appropriate, whereby innovation would be stimulated by outcome- (residue-) directed regulation (Swinton). The potential of information technology may also increasingly be deployed for other purposes. Thus, encouragement for benchmarking of pesticide usage, registration and reporting of usage and support of decision making with relevant tools might more easily be implemented (De Snoo).

Because of such complications some would argue that a more prudent approach is to implement non-chemical routes for pest control. However, the agro-ecological knowledge required for such a strategy is far from developed, as Struik & Kropff make very clear.

Socio-economic institutions

Regulation has been brought forward as a major area of concern in discussions oriented to the socio-economic sphere. It therefore emerges as a focal issue for solutions with far-reaching consequences for the processes in the two other spheres and the interaction between spheres.

One point of particular concern that is supported by the analysis in different chapters is the notion that the current system is sharply bounded and rather inflexible in nature. With regard to registration practices it can be noted that over and on top of the bottlenecks identified by Vogelezang-Stoute – arrears in re-registration, difficulty in dealing with cumulative effects, and substitution – the added purpose of protecting the environment to the largest extent possible is not yet taken fully into account. For instance, Parris & Yokoi argue that the risk-based admission and regulation of pesticides is based on the assertion that cost-benefit assessments are possible. However, such assessments are very hard to make in practice. On the one hand, the OECD attempts to address the pressing lack of information required by increasingly sophisticated risk management protocols, though its attempts need to be followed by member states. On the other hand, pesticide policies will need simple guidelines on how to improve the admission and management processes. Furthermore, utilisation of scientific research and knowledge in a more complete picture of risks in assessment stages could provide advantages over dossier-based admission procedures (Govers *et al.*, Chapter 9). There are several uncertainties in the testing protocols and subsequent authorisation decision making. There is limited consideration of the toxic and other properties of metabolites formed in biotic and abiotic pathways, and no specific consideration of impurities such as products of unwanted side or subsequent reactions. Especially in the case of chlorinated compounds, the reactivity of chlorine could be a source of unexpected and unwanted surprises. Moreover, the results of the monitoring of pesticide concentrations in various environmental compartments rarely feed back into decision making on pesticide authorisation. The point that there is a need for continuous readjustment between the authorisation of chemicals and the monitoring of their fate in the environment has also been stressed in other contributions. In Chapter 3, for example, it is stated that the extensive juridical fights necessary to correct apparent mistakes with regard to the admission of chemicals require readjustment of admission and control phases. The regulatory system seems to be in the need of correction with regard to the possibility of withdrawing active ingredients from the market.

Parallel to inflexibility in the regulatory system is the predominant orientation of agrochemical innovation to authorisation policies. An important actor in agrochemical innovation is the agrochemical industry, which claims that constant innovation in chemical pesticides is essential for a correct solution to the complex problem of crop protection, yield insurance and pest resistance. It is suggested that breakthrough technologies and new compounds will emerge from the innovation process. However, den Hond (Chapter 4) demonstrates that there is a strong case for similarity, instead of variation of efforts, of large agrochemical companies and much less breakthrough innovation than the rhetoric would suggest. The innovator works within a much more restrictive framework than suggested by this broad mission. One element of this context is the status of the market. The pesticide market is largely a saturated market, restraining the options for incumbents to new product development. An example of such restraining factors is the existing complex of actors that function together: the farmer is accustomed to applying chemicals in certified routine procedures, the chemical industry is used to selling active ingredients and not bothering too much with other methods. Another element is the status of pesticide authorisation. Initially, adverse health effects were the main focus of pesticide authorisation; recently environmental concerns have been added. The search process for new pesticide products has consequently been optimised to satisfy regulatory demands on environmental concerns as it has been made operational in testing protocols.

Secondly, a major characteristic of the regulatory sub-system is the political character of changes that are imposed on it. At first sight, the interaction between legal intricacies and policy would seem rather straightforward. However, as has become clear in Chapters 3 and 5, the actual system of rules is distinct from the practices of both political and agricultural decision making on the application of pesticide products. Thus, Vogelezang-Stoute concludes that there is a demonstrable difference between the content of the rules that make up the regulation and the actual decision making around the availability of pesticide products in the market. Likewise, de Snoo demostrates differences between the regulations on pesticide applications and actual pesticide applications. This requires a constant readjustment of the regulatory system that sometimes takes a back seat to overarching policies such as those of EC harmonisation. Moreover, the regulatory system itself, the translation of the various concerns around pesticides into legal requirements to their application, may be far from adequate. Irwin and Rothstein (Chapter 5) argued that 'regulatory science' is influenced by regulatory pressures, the large role of private R&D activity, and the changing nature of science itself. Their discussion suggests a growing awareness that regulatory science is engraved in institutional practices that to a large extent encompass non-scientific considerations. It can better be considered a process in which various actors – industry scientists, test laboratories, pesticide authorities – play their role. In the description of steps taken in the UK, such as the Pesticides Forum, an increased awareness of this character of regulatory science can clearly be seen. One of the challenges necessary for continued public support for this system of decision making is the need to transfer implicit and tacit understandings of the economic, legislative, agricultural and scientific uncertainties and intricacies of the decision making to the broader interested public.

A third, and final, aspect is the legitimisation of regulatory statutes and assessment of active ingredients by scientific evidences and the adamant role of experts. Vogelezang-Stoute (Chapter 3) argues that the broad use of procedural renewal demonstrates that many substances are not tested in the light of *current* scientific knowledge. Govers *et al.* argue that other questions and uncertainties may be relevant too. In that light the authorisation procedures often do not reach the goal of proper decision making based on most relevant scientific evidence. The role of scientists as described by Irwin and Rothstein is particularly relevant. The suggested underlying pattern is that regulatory communities are quite closely knit and cross-cut the borders between public and corporate science. In other words, the social practices that constitute the regulatory and registration processes may be based in communities that lack self-control. Increasing their flexibility may be one way of approaching this problem (Govers *et al.*). Ironically, another but probably unintended outcome of EU harmonisation might be a widening of the community and increased attention to proper application of procedures. Thus, the conclusion of Vogelezang-Stoute may be good to note as a last point. Positive effects can be expected from the further implementation of the EU directive in national policies. Furthermore we would argue that, on the basis of a clear political vision on the further development of regulation and sufficient pressure from stakeholder groups to incorporate considerations of integrated control in the implementation of the EU directives, sustainability goals may become operational in regulatory policies. Reasserting the role of advanced science in regulatory decision making would require science and regulation to be more exposed to policy priorities and the influence of a broader group of stakeholders.

Predominant in our attention to socio-economic institutions has been the regulatory context, notably the authorisation process of active ingredients and pesticide products. However, agricultural production and innovation are also intertwined with other developments in the socio-economics sphere that warrant attention with regard to solutions to the pesticide problem. Two developments in society that illustrate this interconnection have been discussed. One is the level of public support for particular forms of agricultural production and innovation; the other is the increasing importance of international agreements as disciplining regimes for international trade and national agricultural policies.

In several ways, genetically modified (GM) crops and food products are not so radically divergent from previous practices of breeding and improvement of food production with the help of microbiological processes. Nevertheless, the idea of direct manipulation of genetic material has led to a sometimes confusing discussion between proponents and opposition (Carr, Chapter 11). This area of conflict is reinforced by a decreasing societal sympathy for industrialised food production. Animal disease epidemics and the way these have been treated in the political arena within the European Union have made the situation even more complicated. One possible outcome of the different discussions is increased attention to the regional and controlled character of food production (Barling, Chapter 13). Of course, this political momentum may only be temporary; however, it also opens up potential for reinforcing existing shifts to less intensive and chemically supported ways of producing. For instance, van der Grijp (Chapter 12) analyses, how along with these shifting social preferences, food and retail industries are reinforcing their control on agricultural production. In the discussion on biological farming and ICM one of the problems that was brought forward concerns the development of a consistent supply of food produced under organic or ICM controls. The perspective for further development lies in the consistent drive from the retailers to develop markets for organic products. Retailers are increasingly determining the shape of the food sector; power in the food chain is shifting from farmers and food production to distribution and retail. For instance, the concentration of market power in a small number of very large retailers, as is currently occurring in northwestern Europe, puts much of the market power and responsibility in only a few hands. The initiatives described by van der Grijp will reinforce the drive towards more sustainable (and accountable) agricultural practices. For farmers the development of consistent sets of requirements on the demand side may convince them to analyse, document and improve their production methods. Such action will reinforce the pressure on farmers to incorporate social costs in their decision making (*cf.* Chapters 6 and 7).

Both the supra-national and intergovernmental regimes of CAP and WTO provide systems of rules and norms. Such institutional systems are the outcome of negotiations where not always are the intended outcomes ensured. Nevertheless, they influence the conditions under which conventional and alternative agricultural systems of production have to operate. For organic farming, Barling sketches the multifaceted consequences of changes in the institutional system. Organic farming plays out two roles in the system, one of which is the explication of an extensification strategy. The other role is that of a provider of wider public goods, notably rural development and nature conservation. The increasing spread of organic farming takes place at the same time as the introduction of field trials for genetically modified crops. Direct conflict occurs between these alternatives.

Cross-cutting outlook

When we link the different themes that emerge from the discussion of the chapters in a different manner there are two cross-cutting issues: first, the role of stakeholder groups, and second, the dilemmas that result from continuing existing trajectories. Both of these issues were (implicitly or explicitly) left out of subsequent analysis and conclusion, but are highly relevant for the further debate and research on the persistence of the pesticide problem.

The number of different stakeholder groups that are actively involved in discussions on pesticide problems has increased. For instance, increased representation is visible in the role of environmental groups that present their demands to farmers, as well as to the food production chains, notably food retailers. These stakeholder groups build alliances of practice with groups of farmers that want to develop better-adjusted or totally different methods of farming. In a way, these groups seem to exemplify that the paradigm of what is good farming is shifting. This shift is not represented very well as yet in the government sphere of regulation and agricultural policy. In some chapters (examples are Govers *et al.* and Irwin & Rothstein) it is argued that this increased presence should be better recognised in the organisation and design of regulatory processes. With regard to agricultural and trade policy, the assessment of non-economic consequences is an issue that is not yet fully on the agenda. The assessment would require the explicit recognition of essential interests and scientific models to incorporate such concerns. As Parris & Yokoi suggest, this is an area of necessary research and activity. If the more challenging suggestions on this point in various chapters were taken seriously, then subsequently, fundamental research could be done into the manner in which stakeholder interests could be incorporated in the policy phases of analysis, formulation, implementation and evaluation. We surmise that the role of (representatives of) various stakeholder groups for each of these different phases would require a careful analysis of the possibility of incorporating their perspectives. But some need for this is demonstrated by the fact that the stakeholder groups and different perspectives already exert influence. They have accumulated both legitimacy, urgency and power: important attributes to their salience (Mitchell *et al.* 1997).

The second issue of importance is the diagnosis that in a number of the configurations that make up the three spheres rigidity seems to be a problem. Rigidity is the flip-side of specialisation. The level of rigidity of a system increases as a necessary corollary of increased specialisation, but paradoxically it is also required for improving results, and, as such, specialisation may inhibit learning at higher levels (March 1991). However, what would precisely be needed is learning at those higher levels, not only by individual actors, but also among groups of actors, not only in the context of solving isolated problems, but also in enhancing knowledge and understanding of the underlying interdependencies, structures and driving forces. We would argue that work in this direction is urgently needed in relation to the development of farming systems that are – in one way or another – better oriented at environmental management while still profitable (chapters of Struik & Kropff, Wossink & de Koeijer, de Snoo and Barling), and in relation to the implementation and further improvement of pesticide authorisation and regulation processes (chapters of Vogelezang-Stoute and Parris & Yokoi).

References

van Dommelen, A. (1999) *Hazard Identification of Agricultural Biotechnology: Finding Relevant Questions.* International Books, Utrecht.

March, J.G. (1991) Exploration and exploitation in organizational learning. *Organization Science*, 2 (1), 71–87.

Mitchell, R.K., Agle, B.R. & Wood, D.J. (1997) Towards a theory of stakeholder identification and salience: Defining the principle of who and what really counts. *Academy of Management Review*, 22 (4), 853–86.

Index